*I dedicate this book to Wang, Yan Jun whose love has inspired me.*

Al Cutter

# CONTENTS

# FOREWORD

Having been in the electrical trade as an electrician and the business manager of Local 456 IBEW for over 50 years, one of the things I have learned is that the electrical trade is forever changing. There has been much advancement in technology in the electrical industry and today the technology does not change at a reasonable pace, it explodes very rapidly. With the rapid development in photovoltaic, wind, and hydrogen energy, the electrician is constantly challenged to keep up with the dynamic technology. For the instructor of the electrical trade finding current, creditable sources of this information can be difficult. This book is written as an introduction to alternative energy for the student, which makes it a great resource for the instructor.

I have known Al Cutter for over 40 years, and he has always found a way to take the mystery out of the technology and make it easier to understand. Al has taken the subject, alternative energy, and put it in a way that an electrician, student, or engineer can understand. In his book, *The Electrician's Green Handbook,* Mr. Cutter has done a great service for the education of current and future electricians.

Joseph V. Egan
Business Manager Local 456 IBEW

# PREFACE

Since the beginning of the modern age man has looked for alternative energy—coal as an alternative to wood, petroleum as an alternative to whale oil, alcohol as an alternative to fossil fuels. Man has always been looking for a better source of energy. In 1917 Alexander Graham Bell advocated ethanol from corn as an alternative to coal and oil. As a fuel source became scarce and more expensive an alternative energy was sought and this is still true today.

Today, alternative energy is a catchall term that is used for any energy that does not have the undesirable effects of our current fuels. The undesirable effects from our use of current fuels include the difficulties of radioactive waste disposal from nuclear energy production, and the resource pollution and major contribution to global warming resulting from fossil fuels. Alternative energy could mean photovoltaic cells, hydrogen fuel cells and wind power, to name just a few.

Technology is developing very rapidly in new alternative energy sources.

## INTRODUCTION

The objective of the work is to present the current state of the art of alternative energy technology from the practical point of view. There are many books written from a design or justification view. An electrician writes this book from an installation and service point of view. The information needed to understand the workings of these systems will be presented. I will present the building blocks of the systems as to assist the reader to be able to understand the components of the systems and build to the entire system. I use illustrations and drawings throughout to enable the reader to see the technology in practical terms.

## SAFETY FIRST

It is the responsibility of each person for their own safety and the safety of everyone around him or her. Always follow the Occupational Safety and Health Act (OSHA) safety regulations as they are there to protect you and your coworkers.

It is OSHA's mission to prevent work-related injuries, illness, and death. If you have not been trained on OHSA regulations you should take a course at your local union, community college, or trade school.

When working with alternative energy systems the output voltage is very often high voltage, which can be lethal. This voltage can be well over 600 volts DC when

PV cells are wired in series or in wind-powered systems. Also the voltage may be very low; a 48-volt system can be supplying 600 amps.

Always test the circuit for the presence of voltage; never take anyone's word that the circuit is off. Use an approved lock-out tag-out procedure to ensure that you will be safe.

Never work with a live circuit unless absolutely necessary. If you must work with a live circuit, make sure to follow the OSHA regulations for working with live electrical systems and wear the proper personal protective equipment (PPE).

Throughout the book "Safety First" boxes offer safety tips and cautions.

*Your safety is your responsibility, so be safe.*

# INSTRUCTOR COMPANION WEB SITE

An Instructor Companion Web site containing supplementary material is available. This Web site contains an instructor's guide, test banks, image gallery of text figures, and chapter presentations in PowerPoint®. Contact Delmar, Cengage Learning or your local sales representative to obtain an instructor account.

## Accessing an Instructor Companion Web site from SSO Front Door

1. Go to http://login.cengage.com and log in using the instructor e-mail address and password.
2. Enter author, title, or ISBN in the "Add a title to your bookshelf" search box, then click on the search button.
3. Click "Add to My Bookshelf" to add Instructor Resources.
4. At the Product page, click on the Instructor Companion Web site link.

## New Users

If you are new to http://www.cengage.com and do not have a password, contact your sales representative.

# ABOUT THE AUTHOR

Albert F. Cutter, Sr. is an electrician first. He has traveled around the world installing systems and teaching systems technologies. As a 45-year member of Local 456 IBEW, he has worked on systems of all types and sizes. He has spent much of his time teaching at night school for the IBEW apprenticeship courses and college-level courses on industrial control systems. He designed and installed networked computer systems and taught computer technology while working in China for 3½ years. He is fluent in over 30 computer languages and learning a new one every day. He is a published author and patented inventor with many patents pending. At 63 years old he is still working in the electrical trade. In 2010 he founded AppsDev LLC, which designs and develops applications for the Apple and Android smartphones and tablets. As a photographer he enjoys spending his free time taking nature photographs.

# ACKNOWLEDGMENTS

I want to thank the following for their encouragement and assistance, without which this book would not have been written.

John Megna with SMA USA for his technical assistance

Technical reviewers:
- Jamie Lim
- Pat Lyons
- Miguel de Leon

Editors:
- Stacy Masucci, Acquisitions Editor; Delmar, Cengage Learning
- John Fisher, Senior Product Manager; Delmar, Cengage Learning

The entire book staff from Delmar, Cengage Learning

Special thanks to my Brothers and Sisters of Local 456 IBEW, who helped and encouraged me in writing this work, and to my daughter, Barbara, who has helped me so much.

I would also like to thank the following individuals who reviewed the manuscript in detail:

Thomas Collins, Gateway Community and Technical College, Florence, KY

Thomas Henderson, Tulsa Community College, Tulsa, OK

Greg Skudlarek, Minneapolis Community Technical College, Minneapolis, MN

# CHAPTER 1

# CONVERSION

## objective

This chapter will give the reader the understanding of the conversion of DC to AC and AC to DC, which is at the heart of most alternative energy systems. This section will explain the technology of the rectifier and inverter. It includes the many forms and types of rectifiers and inverters. Included will be the basic single diode to the full wave bridge rectifier and the use of micro-inverters, stand-alone inverters, string inverters, and centralized systems.

The user will also have a working knowledge of the monitoring of these systems and the many topologies used in the field today.

After completing this chapter the user will have an understanding of these rectifiers, inverters, and system topologies.

## WHAT WE NEED TO KNOW

The rectifier converts an alternating current (AC) to a near straight-line waveform direct current (DC). The inverter converts a straight-line waveform to an alternating waveform. Both of these technologies are used in alternative energy systems. It is not only a good practice, it is required that AE systems are monitored to maintain peak efficiency and performance. It is essential to have a good understanding of these systems.

## INTRODUCTION

At the heart of all alternative energy systems is a power conversion device. Typically alternative energy systems output direct current which must be converted to alternating current for use. But as you will see later in the book some produce dirty AC and need to be converted to DC then back to clean AC.

Ever since Thomas Edison discovered the "Edison Effect" on February 13, 1880, which of course he patented, we have been converting waveforms. The Edison Effect was the first thermionic diode (vacuum tube diode). When he was

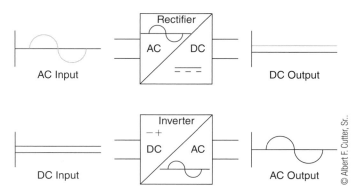

**Figure 1-1** Standard symbols and waveforms for the rectifier and inverter.

1

working on the electric light bulb he noticed that most of the filaments burned out from the positive side. He created a bulb with a metal plate in it and discovered that a current passed through the gas to the plate but only in one direction. It took twenty years before a practical use for the device was discovered.

Back in Edison's time the power that we used for our homes was DC, which was good for that period. But DC was very inefficient for power distribution. It is believed that Nikola Tesla was the inventor of the transformer and his work with George Westinghouse helped to develop the method of AC power distribution, which led to a bitter debate between Edison and Westinghouse known as the "War of the Currents." But when the smoke cleared, Westinghouse won, and today we still have AC grid distribution.

In our everyday life we use many devices that are powered by DC but need AC to operate. Your laptop and your smartphone are just a few. Power conversion is very important in the use of alternative energy, and it is very important that you have an understanding of the methods used to convert waveforms.

# WAVEFORMS

In our investigation of alternative energy (AE) we will be presented with many types of waveforms. They go from the flat line of direct current (DC) to the sinuous waveform of alternating current (AC). There are many waveforms in the conversion of energy in an AE system, like square waves, modified square waves, pulsing DC, and wild AC, to name a few. These waveforms can be seen on an oscilloscope, which is a visual voltmeter. The mentioned waveforms will be covered, as they are relevant. But the understanding of the basic waveforms is relevant here. The simplest of these waveforms is the DC output of a battery.

In Figure 1-2 the DC voltage is represented as a positive voltage. Voltage is measured in respect to your reference point over time. You can say that a given voltage is positive with respect to the negative point or is negative with respect to a positive point.

Figure 1-3 is the original method of DC distribution as developed by Thomas Edison. It was a three-wire distribution system having a positive 110 volts and a negative 110 volts with respect to the neutral conductor.

Figure 1-4 is the sinusoidal AC sine wave. This waveform has a peak in the positive direction, then an equal peak in the negative direction. The rate at which the voltage completes from positive to negative and negative to positive in a given time period is called the frequency, which is measured in Hertz. Hertz (Hz) is a unit of measurement that is defined as the number of complete cycles per second. One Hz

**Figure 1-2** Direct Current waveform.

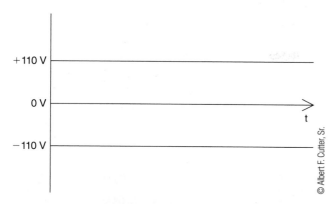

**Figure 1-3** 3-wire DC waveform.

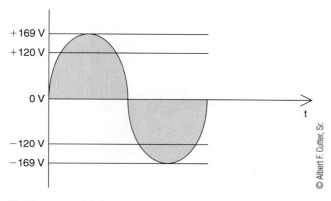

**Figure 1-4** AC waveform.

is equal to one complete cycle per second. In the United States the frequency of distributed AC is 60 cycles or 60 Hz. In most of Europe and Asia the standard AC frequency is 50 Hz.

In Figure 1-4 the peak voltage is 169 volts and the effective voltage is 120 volts AC. The effective voltage is the amount of power required to produce the equivalent heat in a DC circuit, and it is the voltage that we measure with a standard voltmeter. This is called the voltage root mean square (VRMS). This is not the average of the voltage. The VRMS is a mathematical relationship to peak voltage and varies with the type of waveform. In an AC circuit the effective voltage is Vrms = 0.707 × Vpeak or Vpeak = 1.414 × Vrms.

Therefore the AC grid must produce a higher voltage than is used by the load. Most digital meters measure the average of the peak voltage over time and multiply it times 0.707.

# RECTIFIER

A rectifier (Figures 1-5 and 1-6) is an electrical device or component that converts AC to DC. This is called rectification. Rectifiers have many uses as components of power supplies, battery chargers, and detectors of radio waves.

Rectifiers are made up of diodes—these can be solid state or vacuum tubes. The simplest form of rectification is a single diode. This is called half wave rectification. A bridge rectifier uses diodes to provide full wave rectification.

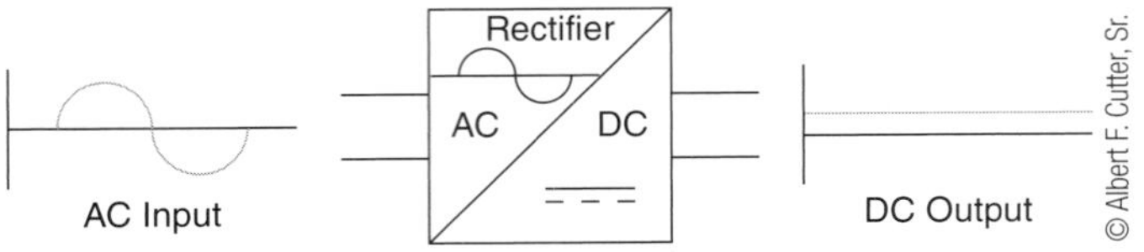

■ **Figure 1-5** Rectifier.

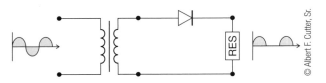

■ **Figure 1-6** Power rectifier.

A rectifier is the opposite of an inverter, which converts DC to AC.

The basic operation of a rectifier is to pass current in one direction only. Benjamin Franklin established that current flows through a circuit from the positive pole (+) to the negative pole (−). Most engineers and electricians follow this theory today. In actuality, free electrons in a conductor almost always flow from negative to positive. In most applications the direction of the current flow is irrelevant. Therefore in the flowing discussion we will use the convention established by Benjamin Franklin.

Figure 1-7 shows the simplest form of rectification—the single diode half wave rectifier. The diode passes the current in only one direction; the positive wave is passed and the negative wave is blocked. Because only half of the AC wave is passed, this type of rectification is not typically used for power transfer.

■ **Figure 1-7** Basic rectifier.

# BRIDGE RECTIFIER

In a bridge rectifier circuit (Figure 1-8), the polarity of the output is the same regardless of the polarity of the input. As the AC waveform goes from positive to negative polarity, the DC circuit remains the same. This is true regardless of the frequency of the AC circuit. This allows it to be used as an efficient power transformer circuit. Because the output polarity of the DC circuit remains the same, it is also used as a device to protect electronic equipment. It can be used to block a DC circuit from an accidental reverse of the input polarity. The bridge rectifier is also called the Graetz circuit after Leo Graetz who invented it in 1880. Figures 1-9 through 1-12 show the flow of current in a bridge rectifier circuit.

In Figure 1-9 when the AC waveform is positive, the current flows on the gray circuit to the positive DC wire and returns on the negative circuit black wire.

In Figure 1-10 when the AC waveform is negative, the current flows on the gray circuit to the positive DC wire and returns on the negative circuit black wire.

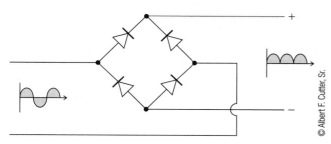

© Albert F. Cutter, Sr.

■ **Figure 1-8** Bridge rectifier.

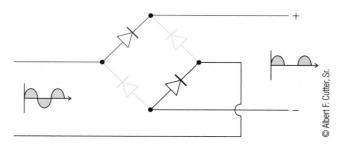

© Albert F. Cutter, Sr.

■ **Figure 1-9** Bridge rectifier current flow.

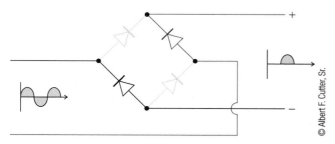

© Albert F. Cutter, Sr.

■ **Figure 1-10** Bridge rectifier current flow.

**Safety First**

Because capacitors store a charge, they can have a lethal charge after the AC source is removed. The larger the capacitor or capacitor bank, the larger the charge. Make sure that there is not a charge before working with them. You should never discharge the capacitor by shorting the terminals; this can expose you to a lethal charge or electrical arc. The circuit should contain a bleeding resistor to drain the capacitor over time.

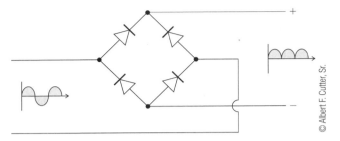

© Albert F. Cutter, Sr.

■ **Figure 1-11** Bridge rectifier.

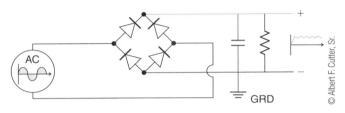

© Albert F. Cutter, Sr.

■ **Figure 1-12** Full wave bridge rectifier circuit.

The bridge circuit, shown in Figure 1-11, is rectifying the complete AC waveform; the output is a ripple DC waveform.

Figure 1-12 shows a full wave bridge rectifier with a filter for smoothing the output waveform. This circuit can pass both positive and negative AC waves; therefore it is the most commonly used rectifier circuit for power transfer. The filter can be a simple capacitor or a bank of capacitors in large systems. The capacitor stores a charge. This charge is discharged when the amplitude of the waveform drops. It is a smoothing capacitor. The output voltage in this circuit is not regulated; when regulated power is required; you must use a power supply with voltage regulation.

# INVERTER

It is impossible to establish with any certainty who was the originator of this commonly used engineering term. However, in 1925 David Prince did publish an article in the GE Review (vol. 28, no. 10, p. 676–81) and cited "The Inverter."

An inverter is an electrical or electro-electrical mechanical device that converts DC to alternating current (AC); the resulting AC can be any required voltage and frequency. In the late nineteenth century to the middle of the twentieth century, DC to AC power conversion was accomplished using motor generator sets (M-G set). A motor generator set is a DC motor driving an AC alternator. Rotary converters were also used, which have a DC motor and an alternator that share a common shaft, field coil, and frame. A commutator is located on one end for the DC motor, and slip rings on the other end for the AC alternator.

Some early inverters used magnetic coils and springs to vibrate or oscillate the DC voltage; these were typically used in car radios. In the early twentieth century, vacuum tubes and gas-filled tubes were used as switches in inverter circuits. The thyratron was the most commonly used type of tube.

In 1957 the solid-state inverter became practical with the introduction of the thyristor or the silicon-controlled rectifier (SCR), and modern solid-state inverters were born.

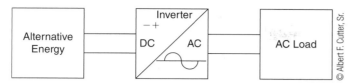

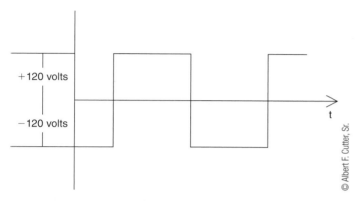

**Figure 1-13** Inverter.

**Figure 1-14** Square wave.

Modern inverters are sophisticated electronic devices. In the following pages I will cover many of the different types of inverters used by alternative energy systems and their use. We will also examine the types of monitoring use. It is important to monitor alternative energy systems to track performance and monitor the current condition of the device. This is necessary to maintain peak performance and maximize energy output and usage.

Most alternative energy (AE) sources produce DC, which must be converted to AC to be used by modern loads as shown in Figure 1-13. Modern appliances and the energy grid require clean frequency-controlled AC. The inverter is supplied with a DC voltage from 0 to 1000 volts and modern inverters efficiently convert this to a sustainable AC power for the load. The AC waveform may be any type and frequency that is required by the load to be served. Inverters are also known as oscillators and are used in radio circuits. For proposes of this chapter we will deal with inverters that have an output frequency of 50/60 cycles.

The simplest waveform to generate is the square wave shown in Figure 1-14. The square wave is low-quality power and is only suitable for resistive loads. Motors and inductive loads are less efficient and will run hotter. Square waves are full of harmonic distortion and should not be used with electronic equipment. With the advent of solid-state electronics, inverters rarely produce square waves and square wave inverters are considered obsolete.

The modified square wave shown in Figure 1-15 is also called quasi square wave or modified sine wave. This waveform is used in low cost stand-alone systems, recreational vehicles (RV), and uninterrupted powers supplies (UPS) for backup power for electronic equipment. As shown, the modified square wave has dead spots between the waveforms to reduce the harmonic distortion. This makes it more suitable for loads than the square wave. Modified square wave is incompatible with many appliances that have electronic controllers. Some other incompatible loads are light dimmers, laser printers, and chargers for cordless tools. Other loads will run less efficiently, and inductive loads like motors will run slower and

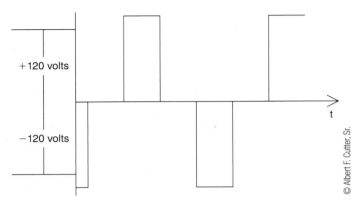

■ **Figure 1-15** Modified square wave.

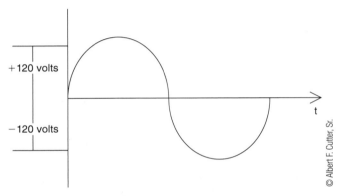

■ **Figure 1-16** Sine wave.

hotter. Modified square wave inverters cannot be connected to the utility grid and with the high cost of alternative energy kilowatts the cost of a pure sine wave inverter will be offset by improved efficiency. Inverters with modified square wave can only be used for stand-alone island systems.

The most commonly used and the most efficient AC waveform is the sine wave shown in Figure 1-16. Sine wave inverters are suitable for all loads that are within the voltage and current output ratings. Their output current has a low harmonic distortion. The total DC component of the inverter output current must be less than 5% to meet the standards for grid connection. Modern inverters with sine wave output meet these requirements and have an efficiency rating over 90%.

## Maximum Power Point Tracker (MPPT)

Output from the AE system will change due to variables of the system. As the sun tracks across the photovoltaic cells, power output changes due to changes in the irradiance level and temperature. The speed of the wind turbine changes due to changes in wind speed. Small changes in the fuel of a fuel cell will cause changes in the cell output. There is a single operating point, at any output current level, where the values of the current (I) and voltage (E) of the AE result in a maximum power output. These values correspond to a particular resistance (R), which is equal to E/I as stated by Ohm's Law (R = E/I). A PV cell has an exponential relationship between current and voltage, and the maximum power point (MPP) occurs at the knee of the curve, where the resistance is equal to the negative of the differential resistance (E/I = −dE/dI). Maximum power point trackers utilize control circuits with logic to search for this point and this allows the inverter circuit to extract the maximum power available from an AE at any output level. Not all inverters have MPPT capability; those that do have a greater efficiency.

## California Energy Commission (CEC) Standards

The California Energy Commission (CEC) was the first state agency to establish standards for AE inverters. The Eighth Edition of the ERP Guidebook can be downloaded from the CEC Web site at http://www.energy.ca.gov.

The CEC Guidebook set forth the standards that are used for AE systems in California.

These procedures are to be used for the testing and evaluation of inverter technology and are considered by most manufacturers to be the de facto standard test procedures.

The following are the links to these test procedures documents:

Performance Test Protocol for Evaluating Inverters Used in Grid-Connected Photovoltaic Systems, prepared by Sandia National Laboratories, Endecon Engineering, BEW Engineering, and Institute for Sustainable Technology, October 14, 2004 version.

http://www.energy.ca.gov

"Guidelines for the Use of the Performance Test Protocol for Evaluating Inverters Used in Grid-Connected Photovoltaic Systems."

http://www.gosolarcalifornia.org

The Institute of Electrical and Electronics Engineers (IEEE) is an international non-profit, professional organization for the advancement of technology related to electricity. It has the most members of any technical professional organization in the world, with more than 350,000 members around the world.

The IEEE has set the following standards forth, which have been adopted by most manufactures.

## IEEE 929-2000

This standard practice contains guidance regarding equipment and functions necessary to ensure compatible operation of PV systems that are connected in parallel with the electric utility. The full recommendation IEEE 929-2000 STD can be found at http://standards.ieee.org/. Copyright © 2008 IEEE.

## IEEE 1547

This standard is for Interconnecting Distributed Resources with Electric Power Systems.

The full recommendation can be found at IEEE 1547. STD can be found at http://grouper.ieee.org/. Copyright © 2008 IEEE.

An inverter that connects to the utility grid is commonly called a grid-tied inverter as shown in Figure 1-17. For an inverter to connect to the utility grid it must have a pure sine wave output and meet the requirements of IEEE Standard 929-2000 and UL Listed 1741. These standards require that output waveform be less than a 5% total harmonic distortion and that the inverter will disconnect if the power is lost on the utility grid. While disconnected the inverter must continue to sample the grid voltage. When the grid voltage has again stabilized and after the required

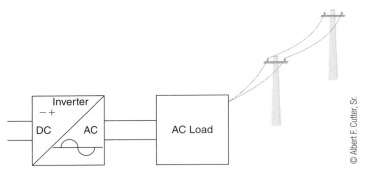

■ **Figure 1-17** Grid-tied inverter.

five-minute delay, the inverter will reconnect to the grid and deliver power from the AE system; this is called anti-islanding. Islanding of an AE system occurs when the section of the utility grid being supplied by the AE system is disconnected. The AE system would be supplying power to the grid section to which it is connected. Unintended islanding is a concern as it poses a hazard to utility equipment, maintenance personnel, and the public.

## Three basic modes of inverter operation

Figure 1-18 shows an inverter in stand-alone mode, also called island system. In this configuration the inverter is supplied from an AE source, with or without battery backup, and delivers power to the AC loads. Stand-alone systems are used where the grid system is not present. Often in stand-alone systems one or more AE sources are used to provide around-the-clock power. Batteries are commonly used in these systems to store energy that can be used when there is no output from the AE system. Systems are under development that use electrolyzers to convert AE energy into hydrogen for use in off-peak periods when the output of the AE system falls below required levels.

The system shown in Figure 1-19 is the hybrid configuration where the AC load is supplied from another generation source when there is insufficient output from the AE source. This source can be anything from wind turbines, engine-driven generators, micro-hydro plants, and others. The hybrid system does not include a connection to the grid distribution system.

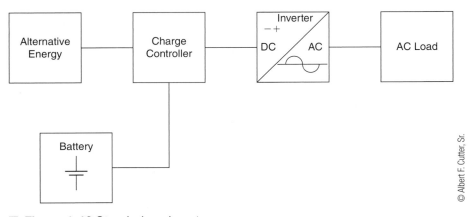

■ **Figure 1-18** Stand-alone inverter.

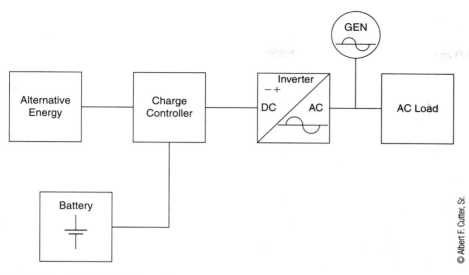

■ **Figure 1-19** Hybrid inverter.

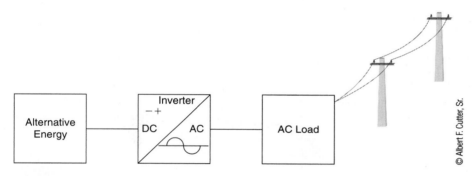

■ **Figure 1-20** Interactive inverter system.

A simple interactive system is shown in Figure 1-20. In this configuration the inverter is fully interactive with the power grid and may supply excess power from the AE system to the power grid. The inverter must be a grid-tied rated inverter to be used in this system. The system can contain one AE source or any number of AE systems to provide power to the AC load and grid distribution system.

## Product Specifications

Different inverter models will have unique specifications. The specifications included here are for a photovoltaic inverter. You should always refer to the manual and specifications before installing or specifying an inverter. Manufacturers will supply you with detailed specifications as requested.

Figures 1-21 and 1-22 show the specifications for the SMA Sunny Boy inverter 5000US / 6000US / 7000US Photovoltaic (PV) Inverter (courtesy of SMA Solar Technology AG). A complete installation manual can be downloaded at http://www.sma-america.com.

The following is a review of the key points of the specification.

**Standard Test Conditions (STC)**—Electrical characteristics of photovoltaic modules vary with irradiance and temperature; the nameplate on the PV module lists the electrical properties of the module under standard test conditions (STC).

## SUNNY BOY 5000US / 6000US / 7000US

> **UL 1741/IEEE 1547 compliant**

> **10 yr. standard warranty**

> **Highest CEC efficiency in its class**

> **Integrated load-break rated AC and DC disconnect switch**

> **Integrated fused series string combiner**

> **Sealed electronics enclosure & Opticool**

> **Comprehensive SMA communications and data collection options**

> **Ideal for residential or commercial applications**

> **Sunny Tower compatible**

# SUNNY BOY 5000US/6000US/7000US
The best in their class

SMA is proud to introduce our new line of inverters updated with our latest technology and designed specifically to meet IEEE 1547 requirements. The Sunny Boy 6000US and Sunny Boy 7000US are also compatible with SMA's new Sunny Tower. Increased efficiency means better performance and shorter payback periods. All three models are field-configurable for positive ground systems making them more versatile than ever. With over 750,000 fielded units, Sunny Boy has become the benchmark for PV inverter performance and reliability throughout the world.

**Figure 1-21** Specifications for the SMA Sunny Boy 5000US / 6000US / 7000US. (See detailed specifications at http://www.sma-america.com.)

## Technical Data
## SUNNY BOY 5000US / 6000US / 7000US

| | SB 5000US | SB 6000US | SB 7000US |
|---|---|---|---|
| Max. Recommended Array Input Power (DC @ STC) | 6250 W | 7500 W | 8750 W |
| Max. DC Voltage | 600 V | 600 V | 600 V |
| Peak Power Tracking Voltage | 250 – 480 V | 250 – 480 V | 250 – 480 V |
| DC Max. Input Current | 21 A | 25 A | 30 A |
| DC Voltage Ripple | < 5% | < 5% | < 5% |
| Number of Fused String Inputs | 4 | 4 | 4 |
| PV Start Voltage | 300 V | 300 V | 300 V |
| AC Nominal Power | 5000 W | 6000 W | 7000 W |
| AC Maximum Output Power | 5000 W | 6000 W | 7000 W |
| AC Maximum Output Current (@ 208, 240, 277 V) | 24 A, 20.8 A, 18 A | 29 A, 25 A, 21.6 A | 34 A, 29 A, 25.3 A |
| AC Nominal Voltage / Range | 183 – 229 V @ 208 V | 183 – 229 V @ 208 V | 183 – 229 V @ 208 V |
| | 211 – 264 V @ 240 V | 211 – 264 V @ 240 V | 211 – 264 V @ 240 V |
| | 244 – 305 V @ 277 V | 244 – 305 V @ 277 V | 244 – 305 V @ 277 V |
| AC Frequency / Range | 60 Hz / 59.3 Hz – 60.5 Hz | 60 Hz / 59.3 Hz – 60.5 Hz | 60 Hz / 59.3 Hz – 60.5 Hz |
| Power Factor | 1 | 1 | 1 |
| Peak Inverter Efficiency | 96.8 % | 97.0 % | 97.1 % |
| CEC weighted Efficiency | 95.5 % @ 208 V | 95.5 % @ 208 V | 95.5 % @ 208 V |
| | 95.5 % @ 240 V | 95.5 % @ 240 V | 96.0 % @ 240 V |
| | 95.5 % @ 277 V | 96.0 % @ 277 V | 96.0 % @ 277 V |
| Dimensions W x H x D in inches | 18.4 x 24.1 x 9.5 | 18.4 x 24.1 x 9.5 | 18.4 x 24.1 x 9.5 |
| Weight / Shipping Weight | 143 lbs / 154 lbs | 143 lbs / 154 lbs | 143 lbs / 154 lbs |
| Ambient temperature range | –13 to +113 °F | –13 to +113 °F | –13 to +113 °F |
| Power Consumption: standby / nighttime | < 7 W / 0.25 W | < 7 W / 0.25 W | < 7 W / 0.25 W |
| Topology | PWM, true sinewave, current source | PWM, true sinewave, current source | PWM, true sinewave, current source |
| Cooling Concept | Convection with regulated fan cooling | Convection with regulated fan cooling | Convection with regulated fan cooling |
| Mounting Location Indoor / Outdoor (NEMA 3R) | ●/● | ●/● | ●/● |
| LCD Display | ● | ● | ● |
| Lid Color: aluminum / red / blue / yellow | ●/○/○/○ | ●/○/○/○ | ●/○/○/○ |
| Communication: RS485 / Wireless | ○/○ | ○/○ | ○/○ |
| Warranty: 10-year | ● | ● | ● |
| Compliance: IEEE-929, IEEE-1547, UL 1741, UL 1998, FCC Part 15 A & B | ● | ● | ● |

Specifications for nominal conditions
● Included   ○ Option   – Not available

### Efficiency Curves

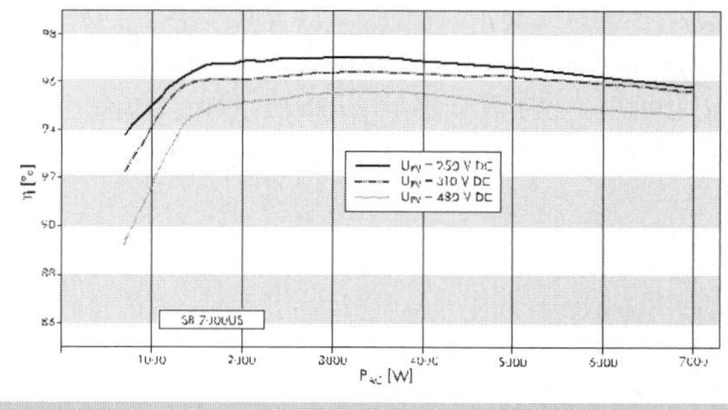

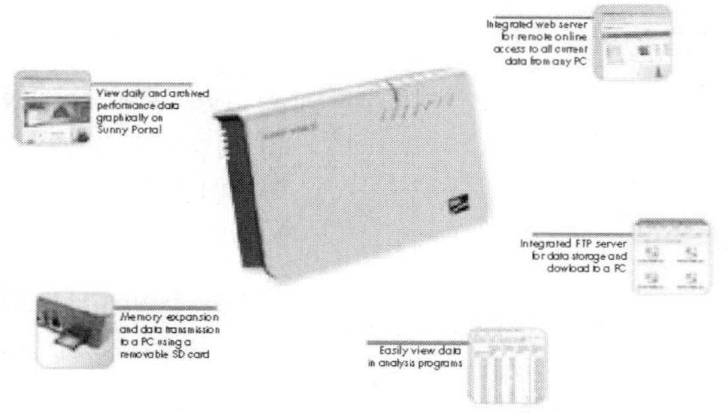

www.SMA-America.com
Phone  916 625 0870
Toll Free  888 4 SMA USA

## SMA America, Inc.

**Figure 1-22** Specifications for the SMA Sunny Boy 5000US / 6000US / 7000US. (See detailed specifications at http://www.sma-america.com.)

This requires that the cell or modules have a temperature of 25°C, an air mass of 1.5, and an irradiance level under 1 kW/m². These conditions are designed for testing in a manufacturing environment but tend to overestimate actual performance, as the cell temperature is rarely at a temperature of 25°C and an irradiance of 1 kW/m² at the same time. This standard is designed as a benchmark or base line of the output of the cells or modules under test conditions.

**Recommended Maximum PV Module Power (STC)**—The maximum PV module output under standard test conditions, which because of test conditions, should be the highest output power of the module.

**DC Maximum Voltage**—The maximum DC input voltage for the inverter. The peak DC output voltage of the AE system must be lower than this value.

**Peak Power Tracking Voltage**—The voltage range that the inverter tracks the maximum power point (MPPT).

**DC Maximum Input Current**—The maximum DC input current of the inverter. The peak DC output current of the AE system must be lower than this value.

**DC Voltage Ripple**—This value is the DC component of the output AC and must be less than 5% for IEEE 929-2000 and UL 1741 compliance.

**Number of Fused String Inputs**—The number of fused terminals for DC input. They will be connected in parallel and will vary from inverter to inverter.

**PV Start Voltage**—The minimum voltage to start the inverter.

**AC Nominal Power**—The power output of the inverter regardless of the AC output voltage setting expressed in watts.

**AC Maximum Output Power**—The maximum output power in watts for any voltage setting.

**AC Maximum Output Current**—The maximum AC current output for a selected voltage setting.

**AC Nominal Voltage Range**—The output voltage range for a given selection; for example, for 240 volts AC, the output range will be between 211 and 264 AC.

**AC Frequency: nominal/range**—This should be 60 Hz for United States systems or within a range of $+/-1$ Hz.

**Power Factor (Nominal)**—Power factor of an inverter is defined as the ratio of the real power flowing to the load to the apparent power—the power measured at the output. This is a number between 0 and 1; the closer to 1, the greater the efficiency of the inverter.

**Peak Inverter Efficiency**—This is the efficiency that is measured by the manufacturer under laboratory conditions.

**CEC Weighted Efficiency**—The California Energy Commission was the first state agency to establish standards for inverters. These standards require that the inverter be tested in real use conditions and are considered to be the most complete standards in the country.

**Compliance IEEE-929, IEEE-1547, UL 1741, UL 1998, FCC Part 15 A and B**—The inverter must meet or exceed the requirements of IEEE-1547 and UL 1741 to be suitable for the inverter to connect to the grid.

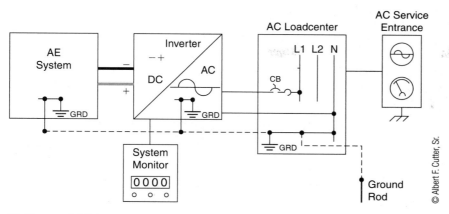

■ **Figure 1-23** Inverter system diagram.

In the Figure 1-23 the AE system output is direct current (DC) connected to the inverter. The DC power is generated by the photovoltaic cells, fuel cells, wind, or any other source. The inverter converts the DC output from the AE system to feed the alternating current to the loadcenter and the utility grid. The requirements of the NEC 2011 code are different for the system depending on each type of AE source to which they are connected. For this reason NEC 2011 code requirements will be covered in the next chapters for each AE system.

### Wire, Fuse, Circuit Breaker, and Disconnect Sizing

The calculation of the wire size for the purposes of this system is 125% of the continuous current rating of the system. For a system having a current of 5.88, the corrected current would be 7.35 amps (1.25 × 5.88 = 7.35 amps). This can be rounded up to the next standard fuse size, which is 10 (amps). The #14 AWG THWN-2 will be adequately protected by a 10-amp fuse.

> **NOTE:** The continuous current rating of a PV system is calculated differently (please see Chapter 3 for those calculations).
> See NEC 2011 Article 240 Overcurrent Protection, which covers the protection of the conductors and disconnects.

## Voltage Drop

Voltage drop must to be taken into account when using a high power DC system or where long cable runs are used. This is the main argument that George Westinghouse used to convince the power companies to switch from DC to AC for the Niagara Falls power plant. Transmission of the power at high voltage and low current with long line transmission power distribution was possible. With the high cost of AE power generation, it is wasteful to dissipate energy to heat on the wires. The cost of larger wires is usually minimal compared to the loss of the AE energy. AE systems with excessive voltage drop are inefficient and will perform poorly.

As stated in NEC 2011 210.19(A)(1) FPN No. 4: Conductors in branch circuits must be sized to ensure a total voltage drop of 3% or less.

Assumptions:

V = voltage

$V_d$ = voltage drop (*in volts, not percent*)

$V_{nom}$ = voltage normal

$\%V_d$ = percent of voltage drop expressed in percent

I = current

Cmils = circular area of a conductor (*NEC 2011 Chapter 9 Table 8/table 1-1*)

L = one-way length of the circuit or feeder

P = watts

K = resistance of conductor (circular mil − ohms/ft.)

= 12 for circuits loaded to more than 50% of allowable circuit capacity copper conductors

= 11 for circuits loaded to less than 50% of allowable circuit capacity copper conductors

= 18 for aluminum or copper-clad aluminum conductors

For the calculation, the system has the following values:

P = 2500 watts

V = 425 volts DC at peak output

L = 500 feet

K = 12 (copper conductors)

Cmil = 4110 (#14 AWG) from table 1-1

As stated in Ohm's Law:

$V_d = I \times R$

$P = V \times I$

$I = P / V$

$\%V_d = \dfrac{V_d \times 100}{V_{nom}}$

First we need to find the DC current at peak output:

$I = P / V$

$I = 2500 / 425$

$I = 5.88$ amperes

| Wire Size | Cmil Area (ohms/ft.) |
|-----------|----------------------|
| #14 | 4110 |
| #12 | 6530 |
| #10 | 10,380 |
| #8 | 16,510 |
| #6 | 26,240 |
| #4 | 41,740 |
| #3 | 52,620 |
| #2 | 66,360 |
| #1 | 83,690 |
| 1/0 | 105,600 |
| 2/0 | 133,100 |
| 3/0 | 167,800 |
| 4/0 | 211,600 |

© Albert F. Cutter, Sr.

**Table 1-1** Wire size.

Calculate the $V_d$ at peak output:

$$V_d = \frac{2 \times K \times L \times I}{Cmil}$$

$$V_d = \frac{2 \times 12 \times 500 \times 5.88}{4110}$$

$$V_d = 17.7$$

Calculate the percentage of voltage drop; the target for the feeders is 1–3%:

$$\%V_d = \frac{V_d \times 100}{V_{nom}}$$

$$\%V_d = \frac{17.7 \times 100}{425}$$

$$\%V_d = 4.04\%$$

Clearly 4.40% is too high; recalculate using #12 wire:

$$V_d = \frac{2 \times 12 \times 500 \times 5.88}{6530}$$

$$V_d = 10.81$$

$$\%V_d = \frac{V_d \times 100}{V_{nom}}$$

$$\%V_d = \frac{10.81 \times 100}{425}$$

$$\%V_d = 2.54\%$$

$\%V_d$ of 2.54% is still a little high; recalculate using #10 wire:

$$V_d = \frac{2 \times 12 \times 500 \times 5.88}{10380}$$

$$V_d = 6.8$$

$$\%V_d = \frac{V_d \times 100}{V_{nom}}$$

$$\%V_d = \frac{6.8 \times 100}{425}$$

$$\%V_d = 1.60\%$$

This is expected and well within the target range of 1–3%. The size of the circuit conductor using the current I = 5.88 #14 AWG wire would have been adequate by the NEC 2011, but once we factor the voltage drop, the correct wire size is #10 AWG.

## String Inverters

Inverters can be connected in parallel to form a string configuration (Figure 1-24). This configuration is used to reduce the length of the DC conductors. Voltage drop, safety, and NEC code requirement are a primary concern when using DC circuits.

The combiner box is mounted close to the AE source to reduce the length of the low voltage DC circuits. This will greatly reduce the effect of voltage drop and the size/cost of the circuit conductors. The inverter can be mounted in a suitable location as close as possible to the combiner box to reduce the length of the DC circuits.

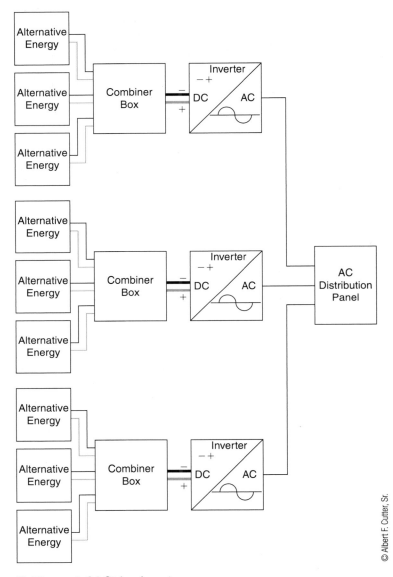

© Albert F. Cutter, Sr.

■ **Figure 1-24** String inverter.

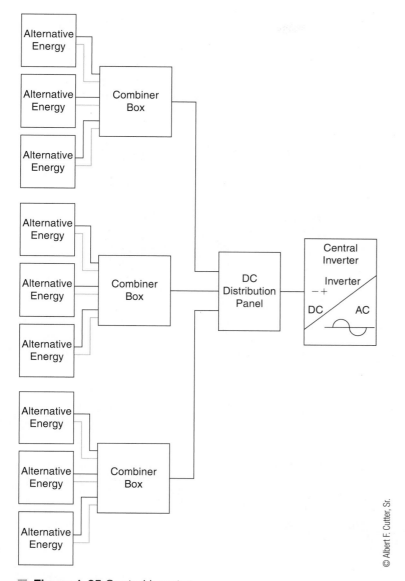

© Albert F. Cutter, Sr.

■ **Figure 1-25** Central inverter.

## Central Inverter

A single large inverter can be used to form a central inverter configuration (Figure 1-25). This configuration has some advantages and disadvantages over the string configuration.

Main advantages: Less equipment to maintain and service. Single point of data collection and monitoring requires less monitoring equipment.

Main disadvantages: Longer DC circuit conductors. Single point of failure and more inputs are combined with less control over system variables.

Figures 1-26 and 1-27 show the specifications for an SMA central inverter system.

## Micro-Inverter

As shown in Figure 1-28, micro-inverters are used on photovoltaic systems where there is an inverter on each PV Module. The inverters take the DC output of the module and convert it to AC at the module. Using MPPT on each inverter allows

SUNNY CENTRAL 250U / 500U

> 97% CEC weighted efficiency

> Integrated isolation transformer

> Graphical LCD interface

> Sunny WebBox compatible

> Optional combiner boxes

> Install indoors or out

> UL 1741 / IEEE-1547 compliant

# SUNNY CENTRAL 250U / 500U
## The ideal inverters for large scale PV power systems

The new Sunny Centrals have integrated isolation transformers and deliver the highest efficiencies available for large PV inverters. A completely updated user interface features a large LCD that provides a graphical view of the daily plant production as well as the status of the inverter and the utility grid. With the optional Sunny WebBox, users can now choose from either RS485 or Ethernet based communications. Designed for easy installation, operation and performance monitoring, the new Sunny Central is the ideal choice for your large scale PV project.

■ **Figure 1-26** Central inverter. (See detailed specifications at http://www.sma-america.com.)

## Technical Data
## SUNNY CENTRAL 250U / 500U

| | SC 250U | SC 500U |
|---|---|---|
| Inverter Technology | True sine wave, high frequency PWM with galvanic isolation | True sine wave, high frequency PWM with galvanic isolation |
| AC Power Output (Nominal) | 250 kW | 500 kW |
| AC Voltage (Nominal) | 480 $V_{AC}$ WYE | 480 $V_{AC}$ WYE / $\Delta$ |
| AC Frequency (Nominal) | 60 Hz | 60 Hz |
| Current THD | < 5% | < 5% |
| Power Factor (Nominal) | > 0.99 | > 0.99 |
| AC Output Current Limit | 300 $A_{AC}$ @ 480 $V_{AC}$ | 600 $A_{AC}$ (@ 480 $V_{AC}$) |
| DC Input Voltage Range | 300 – 600 $V_{DC}$ | 300 – 600 $V_{DC}$ |
| MPP Tracking | 300 – 600 $V_{DC}$ | 300 – 600 $V_{DC}$ |
| PV Start Voltage (Configurable from 300 – 600 $V_{DC}$) | 400 $V_{DC}$ | 400 $V_{DC}$ |
| Maximum DC Current | 800 $A_{DC}$ | 1600 $A_{DC}$ |
| Peak Efficiency | 97.5% | 97.5% (estimated) |
| CEC Weighted Efficiency | 97% | 97% (estimated) |
| Power Consumption | 69 W Standby, <1000 W with fans | 69 W Standby, <1500 W with fans |
| Ambient Operating Temperature | −13 to 113 °F at full power output up to 122 °F at reduced power | −13 to 113 °F at full power output up to 122 °F at reduced power |
| Cooling | Variable-speed forced air | Variable-speed forced air |
| Enclosure | NEMA 3R | NEMA 3R |
| Dimensions: W x H x D in inches | 110 x 80 x 33 | 142 x 80 x 37 |
| Weight | 4200 lbs | 6725 lbs |
| Compliance | UL 1741, IEEE-1547 | UL 1741, IEEE-1547 (pending) |

### Optional
### Sunny WebBox

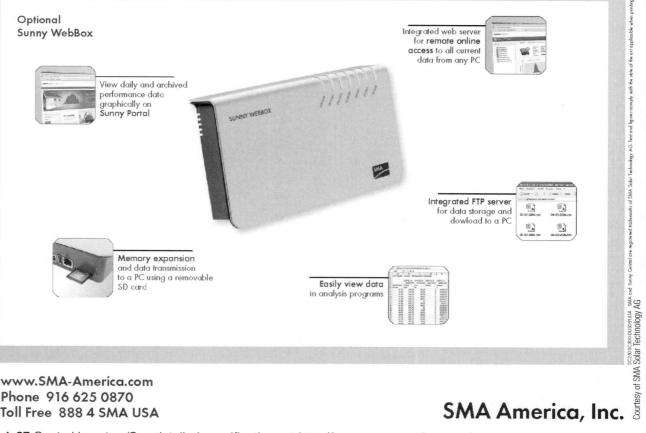

View daily and archived performance data graphically on **Sunny Portal**

Integrated web server for **remote online access** to all current data from any PC

**Integrated FTP server** for data storage and dowload to a PC

**Memory expansion** and data transmission to a PC using a removable SD card

**Easily view data** in analysis programs

**www.SMA-America.com**
**Phone 916 625 0870**
**Toll Free 888 4 SMA USA**

## SMA America, Inc.

■ **Figure 1-27** Central inverter. (See detailed specifications at http://www.sma-america.com.)

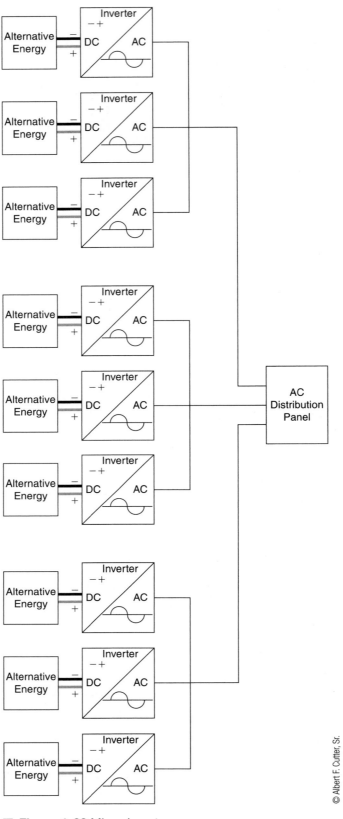

**Figure 1-28** Micro-inverter.

Courtesy of Enphase Energy, Inc.

■ **Figure 1-29** Enphase energy micro-inverter. (See detailed specifications at http://www.enphaseenergy.com.)

for the maximum output from each module. This method also allows for maximum system efficiency. In a string or central inverter system there are usually PV modules in series. This means that the series string will only perform at the level of the weakest module. With micro-inverters each module performs at maximum efficiency. The key advantages to micro-inverters are that you can monitor each module and inverter. This will allow you to identify modules that are not operating well. Different types and sizes of modules can be used in one system. By distributing the inverters there is a reduction in the DC wiring, and combiner boxes and DC disconnects are not required for an installation. The disadvantage is the increase in cost of the inverter modules and their installation. Micro-inverters made by Enphase use power line communication (PLC) to monitor each inverter module (Figure 1-29).

# ALTERNATIVE ENERGY MONITORING SYSTEMS

It is very important to monitor the AE system. Small changes in the system output will affect overall system efficiency. Each inverter type, manufacturer, and model may have different types of systems for monitoring the inverter. These range from simple watt metering to sophisticated computer systems. The SMA Sunny Boy 5000US / 6000US / 7000US Photovoltaic (PV) inverter has the Bluetooth, RS-485, and Ethernet topologies for networking.

## Bluetooth

Figure 1-30 is a typical Bluetooth network. Bluetooth is an open wireless protocol for exchanging data over short distances; the distance depends on the environment. The SMA system uses a remote receiver antenna to increase system performance. This creates personal area networks (PANs). It was originally conceived as a wireless alternative to RS-232 data communication. It can connect several devices. Figure 1-31 is the SMA Sunny Beam Bluetooth receiver. This unit can connect with up to four devices.

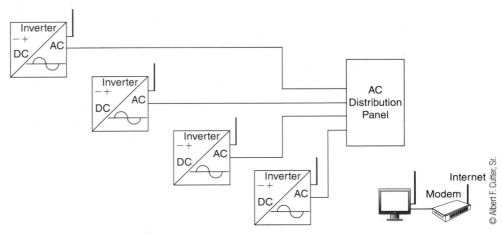

**Figure 1-30** Bluetooth network.

**Figure 1-31** Sunny Beam. (See detailed specifications at http://www.sma-america.com.)

## RS-485

Figure 1-32 is a typical USB Network. RS-485 is the most versatile communication standard in the standard series defined by the EIA. RS-485 connects data terminal equipment (DTE)/remote terminal unit (RTU) directly without the need for modems, which connects several DTEs/RTUs in a network structure. RS-485 has the ability to communicate over long distances, virtually unlimited number of connected nodes with the use of repeaters, and the ability to communicate at faster speeds than RS-232 networks. This is why RS-485 is currently the most widely used communication interface in data acquisition and control applications in which multiple nodes communicate with each other and a central monitoring system.

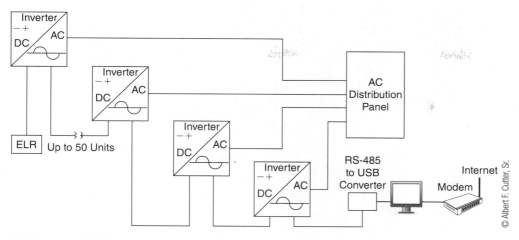

**Figure 1-32** RS-485 network with USB.

With the SMA system up to 50 units can be connected in a multi-point network. This can be up to 4000 feet. As most computers do not have RS-485 inputs, a RS-485 to RS-232 or a RS-485 to USB converter may be needed to connect to the PC. With this method of communication the SMA Sunny Boy products allow the user or installer to transfer valuable system data from the inverters, run system diagnostics, and adjust system parameters.

### Universal Serial Bus (USB)

Figure 1-33 is the B&B Electronics, Inc. RS-485 to USB converter. USB is a serial bus standard for the connecting of devices to a host computer. USB was designed to replace serial, parallel, and Small Computer System Interface (SCSI). It allows many peripherals to be connected using a standardized interface socket and standard drivers. USB standard also allows for the ability to hot swap devices, connecting and disconnecting devices without rebooting the computer or turning off the device. The USB 2.0 specification has a data transfer rate up to 480 Mbit/s.

## Ethernet

Ethernet is a computer networking technology for local area networks (LANs). Ethernet is the most widely used network topology. Typical network speeds are 10/100/1000 Mbit/s. The most commonly used configuration is 10BASET at 10 mbps which runs over four wires (two twisted pairs) on Category 3 or CAT3 cable. CAT5 cable supports 100 BASE-T with speeds up to 100 mbps. CAT6 cable supports 1000 BASE-TX with speeds up to 1000 mbps. A hub or switch sits in the middle of the network and has a port for each node. This is the configuration for all Ethernet configurations regardless of speed. The Ethernet network uses the TCP or TCP/IP protocols. TCP is Transmission Control Protocol and IP is Internet Protocol.

Ethernet is typically used for the backbone of the network. The high speeds of Ethernet are not required for monitoring applications, and the cost of the Ethernet Port Pre-Node network makes it cost prohibitive for the data collection part of the network. It is more costeffective to use RS-232, RS-442, or RS-485 networks than to convert to Ethernet to join the network.

The actual layout of the network will change with the inverter configuration. Figure 1-34 shows one possible layout for a string-configured inverter system. Figure 1-35 shows a converter from DGH Corporation for the inline conversion of RS-485 to Ethernet topology. This is necessary to interface the RS-485 multi-point network to an Ethernet backbone of a network. Figure 1-36 shows an Ethernet network with a central inverter configuration.

**Figure 1-33** RS-485 to USB converter.

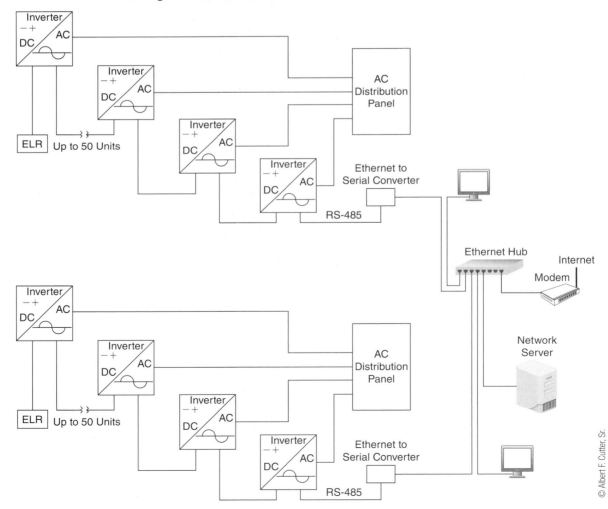

**Figure 1-34** RS-485 / Ethernet inverter string network.

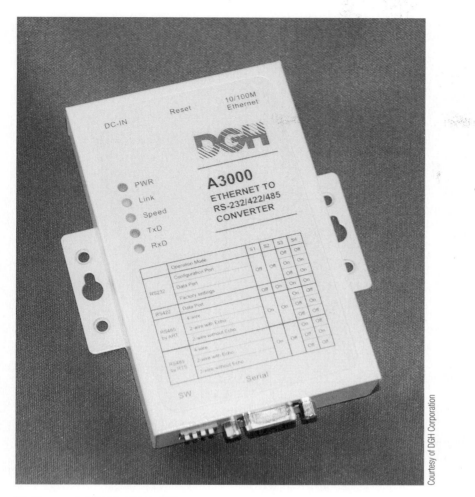

■ **Figure 1-35** RS-485 to Ethernet converter.

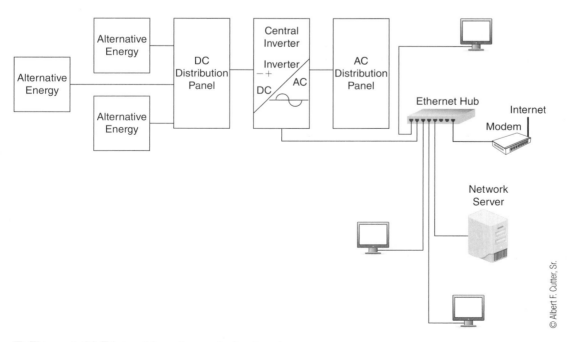

■ **Figure 1-36** Ethernet inverter central network.

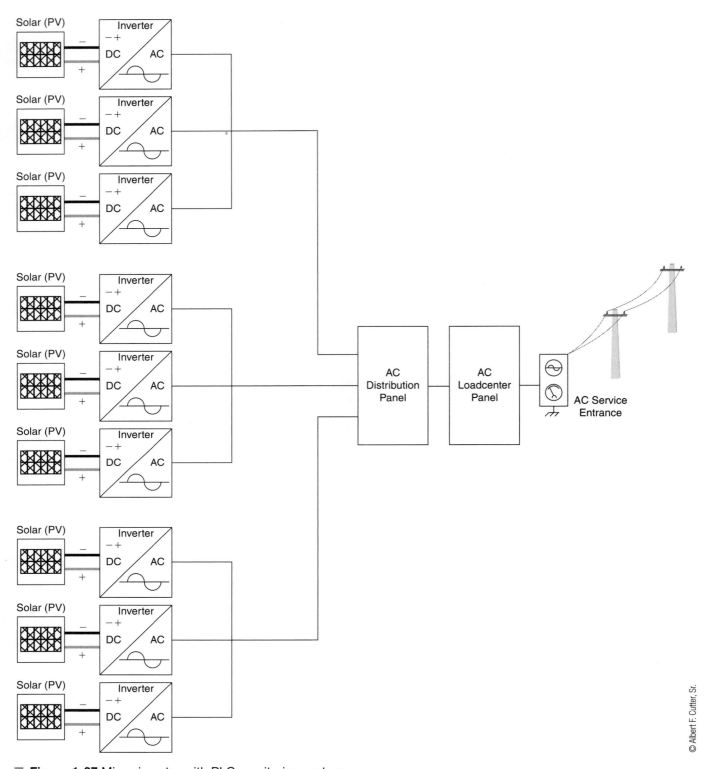

■ **Figure 1-37** Micro-inverter with PLC monitoring system.

# Power Line Carrier (PLC)

Figure 1-37 shows a micro-inverter system with PLC monitoring system. PLC is power line communication or power line carrier system for transmitting data over conductors that also transmit the power line voltage. All PLC systems operate by impressing a modulated signal on the wiring system that is used to supply the power. This is done by imposing a data signal at the zero crossing of the AC sine wave. Power line carrier systems have been around for a long time and are one of the earliest forms of system networking. In the early twentieth century, the street-lights in New York City were controlled by a 500 Hz signal transmitted over the power lines. Today we use PLC to control everything from lights and appliances to electric meters. Most systems use a code that allows 256 devices to communicate on a single system. Systems like the X10 systems, which was developed in 1975 by Pico Electronics in Scotland, allow for the control of any electrical device in the home. Lights can be turned on and off as well as dimmed. In inverter systems, like those from Enphase, Inc., PLC is used to transmit data from each photovoltaic panel to a monitoring system. This allows for control of the system performance, efficiency, and detection of bad cells or inverters. The Enphase micro-inverter allows the user or installer to see the data from each PV cell in the system.

## REVIEW

1. What is the mission statement of OSHA?
2. Draw a DC waveform.
3. Draw an AC waveform.
4. What is the function of a rectifier?
5. What is the function of an inverter?
6. In an AC circuit, if the peak voltage is 339 volts, then what is the effective voltage?
7. Draw the output waveform of a full wave bridge rectifier.
8. The DC component of an inverter output that is connected to the electrical grid must be equal to or less than _____.
    A. 3%
    B. 5%
    C. 2%
    D. 6%
9. The standard for sine wave inverters that connect to the electrical grid must be equal to or greater than _____.
    A. 85%
    B. 100%
    C. 65%
    D. 90%
10. According to the NEC code, what is the allowable percentage of voltage drop in a branch circuit?

# CHAPTER 2

# PHOTOVOLTAIC (SOLAR) SYSTEM INTRODUCTION

## WHAT WE NEED TO KNOW

The reader should have an understanding of basic AC/DC electrical circuits and Ohm's Law.

## INTRODUCTION

The term *photovoltaic* was first used around 1890. The word has two parts: *photo*, from the Greek word for "of light," and *volt*, after the Italian physicist and electricity pioneer Alessandro Volta. Therefore, "photovoltaic" could be literally translated as "of light-electricity." This is the way that photovoltaic materials and devices function—they covert light energy into electrical energy (Figure 2-1).

In 1839 French scientist Edmond Becquerel discovered the photoelectric effect while experimenting with an electrolytic cell, which is made up from two electrodes placed in a conducting solution. Electricity generation increased when the cell was exposed to light.

In 1905 Albert Einstein published his paper on the photoelectric effect (along with a paper on his theory of relativity). Basically this stated that light can be a wave or can be particles, which he called photons. You can measure the waves (frequency) or count the photons but you cannot measure them at the same time.

In 1954 photovoltaic technology was born in the United States when Daryl Chapin, Calvin Fuller, and Gerald Pearson developed the silicon photovoltaic (PV) cell at Bell Labs. This was the first solar cell capable of converting enough of the sun's rays into electrical current to run electrical equipment. Bell Telephone Laboratories produced a silicon solar cell with 4% efficiency and then later they achieved 11% efficiency. As technology has advanced, cells like the Sanyo Heterojunction with Intrinsic Thin Layer (HIT) technology solar cell have achieved a conversion efficiency rating of 23% under test conditions. Also, the National Renewable Energy Laboratory (NREL) has achieved 42.8% conversion efficiency with their triple-junction solar cell under test conditions.

*objective*

This chapter will give the reader an understanding of basic photovoltaic (PV) technology. This chapter will introduce the reader to the history and fundamentals of PV systems. It is important to have an understanding of the components that are used to build PV systems.

■ **Figure 2-1** Photovoltaic materials and devices covert light energy into electrical energy. (See detailed specifications at http://www.sma-america.com.)

Commonly called *solar cells*, individual photovoltaic (PV) cells are electricity-producing devices made of light sensitive semiconductor materials. PV cells come in many types, sizes, and shapes, from smaller than a dime stamp to several inches across. They are connected together to form PV *modules* that can be any shape or size. These modules are assembled and connected to make panels that are combined to form arrays.

With the development of thin-film technology, unlimited shapes are possible. Currently there are manufacturers that make a range of modules, from simple roof coverings to complex shapes that can be installed on roof tiles. Figure 2-2 shows an installation of the thin-film panels on a titled roof.

The only limitation is the shape of the surface that the modules are mounted on and what can be installed easily. PV modules are connected together to form arrays with any shape and power output that is required by the system engineer.

The size of the array is based on the needs of the consumer, the sun's availability at a particular location, and the surface that they are mounted on. The sizes of the arrays are also dependent on the type of inverter used in the power conversion. With the use of the micro-inverter PV, modules of different sizes, shapes, and power output can be combined to form a complete system. Most systems are fixed and mounted on a roof or on the ground with the arrays facing south. But the arrays may be mounted on a tracking system that moves to follow the sun for maximum efficiency.

The panels of the array are the major part of the PV systems. The system is also made up of electrical connections, fuses, disconnects, inverters or electricalizers, and storage systems for use when the sun is not shining. These storage systems are typically battery systems but are now being replaced with hydrogen storage and conversion systems.

Courtesy of Premier Power Renewable Energy, Inc.

■ **Figure 2-2** Thin-film roof panels.

With the improved efficiencies and reduced system cost, PV systems are growing in use worldwide. These can be simple PV modules to power watches and calculators or large arrays used to power entire towns. In the United States there is a large movement to capture the sun as a real alternative energy resource. The state of California is the largest consumer of solar systems with New Jersey a close second. There are current projects in New Jersey that use the sun's rays to produce clean energy. The Federal Express Corporation is installing the largest rooftop array in the North East United States—2.42 mW with 12,400 panels on a single roof. FedEx is installing photovoltaic systems all over the world at their facilities. In the plans are installations in California and Germany this year. PSE&G, a utility company in New Jersey, is putting 200,000 PV panels on utility poles all over the state and rooftop arrays on their buildings in New Jersey. There are major projects all over the world.

In China the Ordos City Government will develop and build a 2-gigawatt solar power plant—the world's largest solar power plant in Inner Mongolia.

The Cimarron I Solar Project is adjacent to the Vermejo Park Ranch in northern New Mexico. It will employ approximately 500,000 photovoltaic modules manufactured by First Solar using its advanced thin-film technology. The project will create over 200 jobs at the construction peak. Electricity generated by the plant will serve a 25-year power purchase agreement with the Tri-State Generation and Transmission Association, a not-for-profit wholesale power supplier to 44 electric cooperatives serving 1.4 million customers across Colorado, Nebraska, New Mexico, and Wyoming.

These are but a few of the current and proposed solar projects.

# PHOTOVOLTAIC DEVICES

Photovoltaic devices can be made from various types of semiconductor materials, deposited or arranged in various structures, to produce solar cells that have optimal performance.

There are three main types of materials used for PV modules. The first type is silicon, which can be used in various forms, including single-crystalline, multicrystalline, and amorphous. The second type is polycrystalline thin-films, with specific discussion of copper indium diselenide (CIS), cadmium telluride (CdTe), and thin-film silicon. Finally, the third type of material is single-crystalline thin-film, focusing especially on cells made with gallium arsenide.

We then discuss the various ways that these materials are arranged to make complete solar devices. The four basic structures include homojunction, heterojunction, p-i-n/n-i-p, and multijunction devices.

# PHOTOVOLTAIC CELL STRUCTURES

The actual structural design of a photovoltaic device depends on the material used in the PV cell and its limitations. The four basic device designs commonly used are:

- Homojunction
- Heterojunction
- p-i-n/n-i-p
- Multijunction

## Homojunction

Crystalline silicon is the primary example of this kind of cell. A single material, crystalline silicon, is doped so that one side is p-type, made up primarily of positive holes, and the other side is n-type, dominated by negative electrons. It is important that the maximum amount of light reaches the junction. The depth of the p/n junction in the cell is very important so that the maximum of useful sunlight is absorbed near it. The free electrons and holes generated by light deep in the silicon diffuse to the p/n junction, and then separate to produce a current if the silicon is of sufficient high quality.

In this homojunction design, several aspects of the cell affect the conversion efficiency:

- Depth of the p/n junction below the surface of the cell
- Amount and distribution of atoms doped on either side of the p/n junction
- Crystallinity and purity of the silicon

Some homojunctions cells have also been designed with the positive and negative electrical contacts on the back of the cell. This design eliminates the shadows caused by the electrical grid on top of the cell. This design of photovoltaic cell has demonstrated an efficiency of 23.4 percent, which is an increase of almost 50 percent efficiency over conventional top grid mounted cells.

# Heterojunction

An example of this type of device structure is a CIS cell, which consists of several layers of differently endowed copper indium diselenide, where the junction is formed by contacting two different semiconductors: cadium sulfide (CdS) and copper indium diselenide (CuInSe$_2$). Thin-film materials that absorb light much better than silicon often use this structure. The top and bottom layers in a heterojunction device have different roles. The top layer, or "window" layer, is a material with a high bandgap selected for its transparency to light. Almost all incident light passes through the window and reaches the bottom layer, which is a material with low bandgap that readily absorbs light. This light then generates electrons and holes very near the junction, which helps to effectively separate the electrons and holes before they can recombine.

Heterojunction devices have an inherent advantage over homojunction devices, which require materials that can be doped both p- and n-type. Many PV materials can be doped either p-type or n-type, but not both. Again, because heterojunctions do not have this constraint, many promising PV materials can be investigated to produce optimal cells.

Also, a high-bandgap window layer reduces the cells' series resistance. The window material can be made highly conductive, and the thickness can be increased without reducing the transmittance of light. As a result, light-generated electrons can easily flow laterally in the window layer to reach an electrical connection.

# p-i-n and n-i-p Devices

Typically, amorphous silicon thin-film cells use a p-i-n structure, whereas cadmium telluride (CdTe) cells use an n-i-p structure. The basic scenario is as follows: A three-layer sandwich is created, with a middle intrinsic (i-type or undoped) layer between an n-type layer and a p-type layer. This geometry sets up an electric field between the p- and n-type regions that stretches across the middle intrinsic resistive region. Light generates free electrons and holes in the intrinsic region, which are then separated by the electric field.

In the p-i-n amorphous silicon (a-Si) cell, the top layer is p-type a-Si, the middle layer is intrinsic silicon, and the bottom layer is n-type a-Si. Amorphous silicon has many atomic-level electrical defects when it is highly conductive. So very little current would flow if an a-Si cell had to depend on diffusion. However, in a p-i-n cell, current flows because the free electrons and holes are generated *within* the influence of an electric field, rather than having to move toward the field.

In a CdTe cell, the device structure is similar to the a-Si cell, except the order of layers is flipped upside down. Specifically, in a typical CdTe cell, the top layer is p-type cadmium sulfide (CdS), the middle layer is intrinsic CdTe, and the bottom layer is n-type zinc telluride (ZnTe).

# Multijunction Devices

This structure, also called a tandem or cascade cell, can achieve higher total conversion efficiency by capturing more of the solar spectrum. In the typical multijunction cell, individual cells with different bandgaps are stacked on top of one another. The individual cells are stacked in such a way that photons fall first on the material having the largest bandgap. Photons not absorbed in the first cell are transmitted to the second cell, which then absorbs the higher-energy portion of the remaining solar radiation while remaining transparent to the lower-energy photons. These selective absorption processes continue through to the final cell, which has the smallest bandgap.

A multijunction device (Figure 2-3) is a stack of individual single-junction cells in descending order of bandgap (Eg). The top cell captures the high-energy photons (Eg1) and passes the rest of the photons on to be absorbed by medium-bandgap cells (Eg2). The remaining photons are passed on to the lower-bandgap cells (Eg3).

A multijunction cell can be made in two different ways. In the mechanical stack approach, two individual solar cells are made independently, one with a high bandgap and one with a lower bandgap. Then the two cells are mechanically stacked, one on top of the other. In the monolithic approach, one complete solar cell is made first, and then the layers for the second cell are grown or deposited directly on the first.

Shown in Figure 2-4 is a multijunction device that has a top cell of gallium indium phosphide, then a "tunnel junction" which allows the flow of electrons between the cells, and a bottom cell of gallium arsenide.

Research today is mostly in multijunction cells with a focus on gallium arsenide as one or all of the component cells. These cells have efficiencies of more than 35% under concentrated sunlight, which is high for PV devices. Other materials studied for multijunction devices are amorphous silicon and copper indium diselenide.

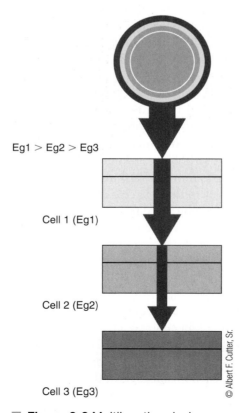

Eg1 > Eg2 > Eg3

Cell 1 (Eg1)

Cell 2 (Eg2)

Cell 3 (Eg3)

© Albert F. Cutter, Sr.

**Figure 2-3** Multijunction device.

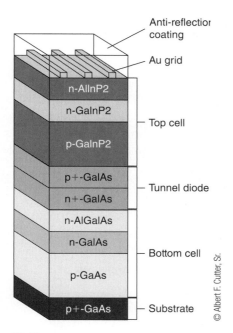

**Figure 2-4** Multijunction device layers.

# PHOTOVOLTAIC CELLS

Figure 2-5 shows a photovoltaic cell-module; the wattages differ depending on the type and size of the cell. For this discussion we will use the specifications for Sanyo Heterojunction with intrinsic thin layer (HIT) solar cells.

Figures 2-6 and 2-7 show the basic structure of a conventional crystalline silicon (c-Si) solar cell and the HIT solar cell. The HIT cell has a conversion efficiency of 23% in the laboratory, which is very high for a solar cell, and a field efficiency of 19.3%.

**Figure 2-5** Photovoltaic cell-module. (See us.sanyo.com for complete specifications.)

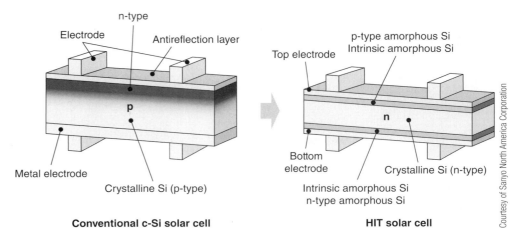

**HIT** (<u>H</u>eterojunction with <u>I</u>ntrinsic <u>T</u>hin Layer) **Solar Cell is composed of thin single-crystalline Si wafer sandwiched by ultra-thin a-Si layers**

Courtesy of Sanyo North America Corporation

**Figure 2-6** Basic structure of a HIT cell. (See us.sanyo.com for complete specifications.)

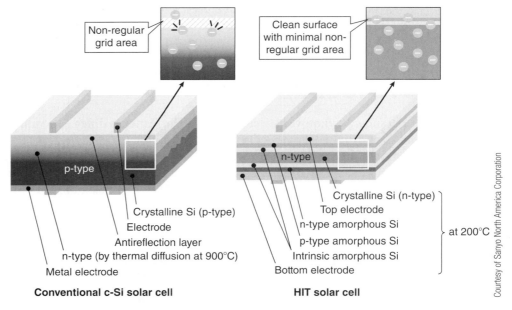

Courtesy of Sanyo North America Corporation

**Figure 2-7** HIT cell structure. (See us.sanyo.com for complete specifications.)

The two most important factors in the conversion of solar energy are the amount of sunlight and temperature. Figure 2-8 shows the efficiency of the solar cell decreases as the cell temperature increases. Advancements in photovoltaic cell technology have minimized the effects of temperature as shown by the chart. The Sanyo HIT cell maintains a higher efficiency over the c-Si cell.

## Ohm's Law

To understand the PV circuits an understanding of Ohm's Law is necessary. Ohm's Law states that $P = E \times I$: where power (P) is measured in watts, voltage (E) is measured in volts, and current (I) is measured in amps. A working knowledge of

**HIT: High Efficiency at High Temperatures**

Excellent feature of temperature dependence:
- High efficiency at high temperatures
- More output power even at high temperatures in summertime

Temperature vs. Conversion Efficiency

Changes in Generated Power Daytime

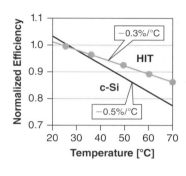

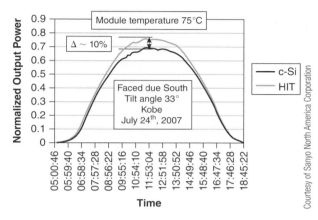

Courtesy of Sanyo North America Corporation

■ **Figure 2-8** HIT efficiency chart. (See us.sanyo.com for complete specifications.)

Ohm's Law is essential to understanding the PV circuits. The three basic formulas for Ohm's Law are $I = E/R$, $R = E/I$ and $E = I \times R$.

**P** = power—measured in watts

**E** = electromotive force (EMF)—measured in volts

**I** = current—measured in amps

**R** = resistance—measured in ohms

Figure 2-9 shows Ohm's Law; we will use this to calculate the system parameters in the following circuit examples. It is tricky to calculate the actual output of the solar panel as it requires measurement at a given irradiation and temperature. For this reason panel nameplates are given for maximum power point under standard test conditions (STC). This requires that the cell or modules have a temperature of 25°C, an air mass of 1.5, and an irradiance level under 1 kW/m². 

The output voltage of a PV cell is at its highest when no load current is being drawn ($I = 0$), this is known as the open-circuit voltage. The output current is at its highest when the cell is short circuited ($E = 0$), this is known as short circuit current. It is important to note that each manufacture and model of solar cells may differ in their physical specifications and electrical characteristics. It is recommended that you always consult the specifications for the units that are used (see Figure 2-12 and Figure 2-13).

Single solar cells produce an open-circuit voltage around 0.5 volts DC and DC currents that range from one to eight amps. Different types of cells will have different output voltages and currents for a given irradiation and temperature. For comparison of equipment the specifications are written using maximum values under standard test conditions. For this discussion we will say that the solar cell has wattage of 2.99 W at STC. With a voltage of 0.5828 volts using Ohm's Law, which states $P = I \times E$, see Figure 2-9, we can calculate the current at $I = P/E$, where the wattage is $P = 2.99$ W and the voltage is $E = 0.5828$. Since, $I = 2.99$ W/0.5828 V, we

**Safety First**

A careful reading of the National Electrical Code is important and is required by anyone working with a PV system. It is very important for the safety of all that close attention is paid to system grounding Article 690.43 and ground fault protection Article 690.5.

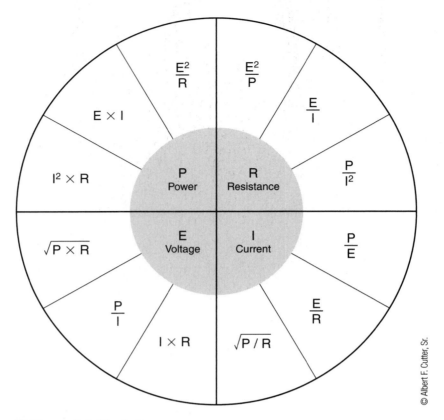

■ **Figure 2-9** Ohm's law.

calculate the current to be I = 5.13. These values are maximum voltages and current, within the standard ratings for solar cells. STC values are used for comparison purpose only. For calculating the circuit conductors you must use the open-circuit voltage and the short circuit current as required by the NEC 2011.

# EXAMPLE CIRCUITS

The circuit in Figure 2-10 shows a single solar cell in a circuit. The blocking diode is used to protect the module from reverse current flow **resulting** from unbalances or surges. A panel will have any number of blocking diodes depending on

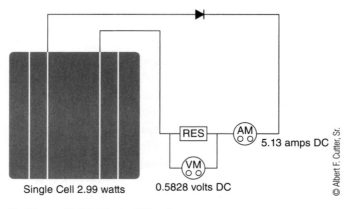

Single Cell 2.99 watts     RES     AM     5.13 amps DC

VM     0.5828 volts DC

■ **Figure 2-10** Single PV cell.

the module configuration and design; see the module specifications of the system for the module configuration.

# PHOTOVOLTAIC PANELS

Solar panels are made up of several cells depending on the manufacturer, types of cells used, the desired output, and physical size requirements ranging from 40 to 96 modules. Panel voltages range from 12 to 60 volts depending on the number of cells and their configuration and characteristics. The high efficiency of the HIT cells (Figure 2-11) allows higher wattages per square foot. As the material and the efficiency improve, the size and number of cells will change.

The Sanyo HIT specifications are shown in Figures 2-12 and 2-13. It is important to note the efficiency of the 215n panel is 17.1%, which is very high for current technology of photovoltaic panels. Because of the high efficiency the system requires fewer panels to produce the required kWh per rated watt of the system. This means less mounting hardware, less space required, and reduced installation costs. Also it is important to note the improved performance at higher temperature range. It is important to become familiar with reading module specifications as they are constantly changing as new technologies evolve.

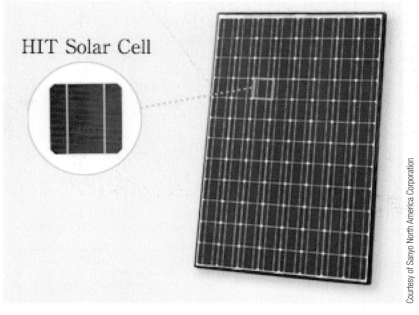

■ **Figure 2-11** HIT HP-205BA3 module 205 watt. (See us.sanyo.com for complete specifications.)

# HIT Photovoltaic Module

**Module Efficiency: 17.1%**
**Cell Efficiency: 19.3%**
**Power Output - 215 Watts**

## SANYO HIT® Solar Cell Structure

p-type/i-type
(Ultra-thin amorphous silicon layer)

Front-side electrode
Rear-side electrode

Thin mono crystalline silicon wafer

i-type/n-type
(Ultra-thin amorphous silicon layer)

### SANYO'S Proprietary Technology

HIT solar cells are hybrids of mono crystalline silicon surrounded by ultra-thin amorphous silicon layers, and are available solely from SANYO.

### High Efficiency

HIT® Power solar panels are leaders in sunlight conversion efficiency. Obtain maximum power within a fixed amount of space. Save money using fewer system attachments and racking materials, and reduce costs by spending less time installing per watt. HIT Power models are ideal for grid-connected solar systems, areas with performance based incentives, and renewable energy credits.

### Power Guarantee

SANYO's power ratings for HIT Power panels guarantee customers receive 100% of the nameplate rated power (or more) at the time of purchase, enabling owners to generate more kWh per rated watt, quicken investments returns, and help realize complete customer satisfaction.

### Temperature Performance

As temperatures rise, HIT Power solar panels produce 10% or more electricity (kWh) than conventional crystalline silicon solar panels at the same temperature.

### Valuable Features

The packing density of the panels reduces transportation, fuel, and storage costs per installed watt.

### Quality Products Made in USA

SANYO silicon wafers located inside HIT solar panels are made in California and Oregon (from October 2009), and the panels are assembled in an ISO 9001 (quality), 14001 (environment), and 18001 (safety) certified factory. Unique eco-packing minimizes cardboard waste at the job site. The panels have a Limited 20-Year Power Output and 5-Year Product Workmanship Warranty.

### Unnecessary Section When Using SANYO

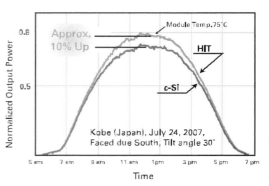

### Increased Performance with SANYO

**Figure 2-12** Specifications for the Sanyo 215n module. (See us.sanyo.com for complete specifications.)

# HIT Power 215N
Photovoltaic Module

## Electrical Specifications

| Model | HIT Power 215N or HIP-215HKHA6 |
|---|---|
| Rated Power (Pmax)[1] | 215 W |
| Maximum Power Voltage (Vpm) | 42.0 V |
| Maximum Power Current (Ipm) | 5.13 A |
| Open Circuit Voltage (Voc) | 51.6 V |
| Short Circuit Current (Isc) | 5.61 A |
| Temperature Coefficient (Pmax) | -0.336%/ °C |
| Temperature Coefficient (Voc) | -0.143 V/ °C |
| Temperature Coefficient (Isc) | 1.96 mA/ °C |
| NOCT | 114.8°F (46°C) |
| CEC PTC Rating | 199.6 W |
| Cell Efficiency | 19.3% |
| Module Efficiency | 17.1% |
| Watts per Ft.[2] | 15.85 W |
| Maximum System Voltage | 600 V |
| Series Fuse Rating | 15 A |
| Warranted Tolerance (-/+) | -0% / +10% |

## Mechanical Specifications

| | |
|---|---|
| Internal Bypass Diodes | 3 Bypass Diodes |
| Module Area | 13.56 Ft² (1.26m²) |
| Weight | 35.3 Lbs. (16kg) |
| Dimensions LxWxH | 62.2x31.4x1.8 in. (1580x798x46 mm) |
| Cable Length +Male/-Female | 40.55/34.6 in. (1030/880 mm) |
| Cable Size / Connector Type | No. 12 AWG / MC4™ Locking Connectors |
| Static Wind / Snow Load | 60PSF (2880Pa) / 39PSF (1867Pa) |
| Pallet Dimensions LxWxH | 63.2x32x72.8 in. (1607x815x1850 mm) |
| Quantity per Pallet / Pallet Weight | 34 pcs./1234.5 Lbs (560 kg) |
| Quantity per 53' Trailer | 952 pcs. |

## Operating Conditions & Safety Ratings

| | |
|---|---|
| Ambient Operating Temperature | -4°F to 115°F (-20°C to 46°C)² |
| Hail Safety Impact Velocity | 1" hailstone (25mm) at 52 mph (23m/s) |
| Fire Safety Classification | Class C |
| Safety & Rating Certifications | UL 1703, cUL, CEC |
| Limited Warranty | 5 Years Workmanship, 20 Years Power Output |

[1]STC: Cell temp. 25°C, AM1.5, 1000W/m² ²Monthly average low and high of the installation site.
Note: Specifications and information above may change without notice.

## Dependence on Temperature

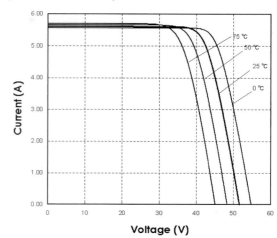

## Dependence on Irradiance

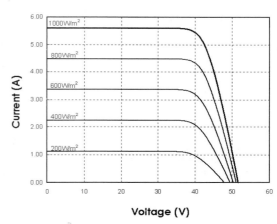

## Dimensions
Unit: inches (mm)

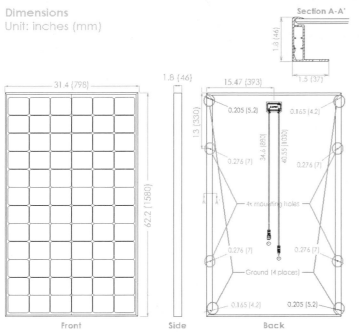

⚠ **CAUTION!**
Read the operating instructions carefully before use of these products

**SANYO**

**SANYO Energy (U.S.A.) Corp.**
A Division of SANYO North America Corporation

550 S. Winchester Blvd., Suite 510
San Jose, CA 95128, U.S.A.
www.sanyo.com/solar
solar@sec.sanyo.com

**Figure 2-13** Specifications for the Sanyo 215n module. (See us.sanyo.com for complete specifications.)

# ELECTRICAL SPECIFICATIONS

**Model**—HIT Power 215n

**Rated Power (Pmax)[1]**—215 W
Rated power at standard test conditions (STC); this is used for comparison only.

**Maximum Power Voltage (Vpm)**—42 V
Maximum power voltage at STC

**Maximum Power Current (Ipm)**—5.13 A
Maximum power current at STC

**Open-Circuit Voltage (Voc)**—51.6 V
According to NEC Article 690.7, the maximum voltage shall be the sum of the open-circuit voltage for all PV modules series-connected with the voltage corrected for the lowest expected ambient temperature. It will be used to calculate circuit maximum system voltage rating for the wire, voltage drop, and all equipment used in the system.

$$Vtc = -0.143 \text{ V/°C}$$

**Short Circuit Current (Isc)**—5.61 A
As required by the NEC Article 690.8 the short circuit current is used to calculate the system equipment.

**Temperature coefficients**—Provide the rate of change (derivative) with respect to temperature of different photovoltaic performance parameters. The derivatives can be determined for short circuit current (Isc), maximum power current (Imp), open circuit voltage (Voc), maximum power voltage (Vmp), and maximum power (Pmax), as well as fill factor (FF) and efficiency ($\eta$). ASTM standard methods for performance testing of cells and modules address only two temperature coefficients, one for current and one for voltage [1, 2]. Outdoor characterization of module and array performance has indicated that four temperature coefficients for Isc, Imp, Voc, and Vmp are necessary and sufficient to accurately model electrical performance for a wide range of operating conditions [3]. ASTM also specifies that temperature coefficients are determined using a standard solar spectral distribution at 1000 W/m² irradiance, but from a practical standpoint they need to be applied at other irradiance levels as well. A variety of practical issues regarding the measurement and application of temperature coefficients still need to be addressed.

**Temperature Coefficient (Pmax)**— $-0.336\%/°C$
The temperature coefficient is the relative change of a physical property when the temperature is changed.

**Temperature Coefficient (Voc)—** $-0.143$ V/°C

NEC Article 690.7 allows this coefficient to be used in place of the value in Table 690.7. When open-circuit voltage temperature coefficients are supplied in the instructions or specification, as shown above, for listed PV modules, they shall be used to calculate the maximum PV open system voltage as required by Article 110.3(B) instead of using Table 690.7 in Article 690.7.

**Temperature Coefficient (Isc)—** 1.96 mA/°C

Isc is the short-circuit current coefficient. It is used as a correction factor for the changes in cell temperature.

**NOCT—** 114.8F (46°C)

**Normal Operating Cell Temperature (NOCT)—** The estimated temperature of a solar PV module when it is operating under 800 W/m² irradiance, 20°C ambient temperature, and a wind speed of 1 m/s. NOCT is used to estimate the nominal operating temperature of a module in the field.

**CEC PTC Rating—** 199.6 W

California Energy Commission (CEC) PTC refers to PVUSA Test Conditions, which were developed to test and compare PV systems as part of the PVUSA (Photovoltaics for Utility Scale Applications) project. PTC is 1000 W/m² solar irradiance, 20°C air temperature, and wind speed of 1 m/s at 10 m above ground level. PV manufacturers use Standard Test Conditions, or STC, to rate their PV products. STC are 1000 W/m² solar irradiance, 25°C cell temperature, air mass equal to 1.5, and ASTM G173-03 standard spectrum. The PTC rating, which is lower than the STC rating, is generally recognized as a more realistic measure of PV output because the test conditions better reflect "real-world" solar and climatic conditions, compared to the STC rating. All ratings in the list are DC (direct current) watts.

**Cell Efficiency—** 19.3%

Efficiency of photovoltaic cell at STC

**Module Efficiency—** 17.1%

Efficiency of photovoltaic module at STC

**Watts per Ft²—** 15.85 W

**Maximum System Voltage—** 600 V

NEC Article 690.7 C states that in a residential installation the photovoltaic system maximum voltage shall be 600 volts.

**Series Fuse Rating—** 15 A

**Warranted Tolerance (−/+)—** $-0\%/+10\%$

## References

[1]. ASTM E 948, "Electrical Performance of Non-Concentrator Terrestrial PV Cells Using Reference Cells."

[2]. ASTM E 1036, "Electrical Performance of Nonconcentrator Terrestrial Photovoltaic Modules and Arrays Using Reference Cells."

[3]. D. L. King, "Photovoltaic Module and Array Performance Characterization Methods for All System Operating Conditions," NREL/SNL Program Review, AIP Press, 1996, pp. 347–368.

# MECHANICAL SPECIFICATIONS

**Internal Bypass Diodes—**3 Bypass Diodes

**Module Area—**13.56 $Ft^2$ (1.26 $m^2$)

**Weight—**35.3 Lbs. (16 kg)

**Dimensions L × W × H—**62.2 × 31.4 × 1.8 in. (1580 × 798 × 46 mm)

**Cable Length +Male/−Female:—**40.55/34.6 in. (1030/880 mm)

**Cable Size / Connector Type:—**No. 12 AWG / MC4™ Locking Connectors

NEC Article 690.8 (B)(1) requires that the circuit conductor and overcurrent devices shall be sized to carry not less than 125% of short circuit current.

The conductor is sized to prevent any significant voltage drop.

MC4TM connectors are guarded, polarized, locking type connectors in accordance with NEC Article 690.33.

**Static Wind / Snow Load:—**60PSF (2880Pa) / 39PSF (1867Pa)

**Pallet Dimensions L × W × H:—**63.2 × 32 × 72.8 in. (1607 × 815 × 1850 mm)

**Quantity per Pallet / Pallet Weight:—**34 pcs / 1234.5 Lbs. (560 kg)

**Quantity per 53' Trailer:—**952 pcs.

# OPERATING CONDITIONS AND SAFETY RATINGS

**Ambient Operating Temperature:—**−4ºF to 115ºF (−20ºC to 46ºC)[2]

**Hail Safety Impact Velocity:—**1″ hailstone (25mm) at 52mph (23m/s)

**Fire Safety Classification:—**Class C

**Safety and Rating Certifications:—**UL 1703, cUL, CEC

■ **Figure 2-14** PV panel Installed on a roof top.

**Limited Warranty:**—5 Years Workmanship, 20 Years Power Output

1 STC: Cell temperature 25c, AM1.5, 1000 W/m$^2$

2 Monthly average low and high of the installation site

The example circuit in Figure 2-15 shows a photovoltaic module with 72 cells with a rated wattage of 215. Photovoltaic cells in the modules are connected in series to produce the required voltage. In series the voltage of the cells is the sum of the cell voltages $E_t = E_1 + E_2 + E_3 + \ldots E_x$. $E_t = 0.5526 * 72$ $E_t = 42$ VDC. The current remains the same $I_t = I_x$ throughout the module. The Sanyo HIT in Figure 2-12 shows a panel made up of 72 solar cells and having wattage of 215 W at a maximum voltage of 42 VDC. We can calculate the current using $I = P/E$, see Figure 2-13. If P = 215 W and E = 42 V, then I = 215 W/42 V, therefore I = 5.13A. These calculations are STC values for comparison only. The actual values will depend on irradiation and temperature.

# PHOTOVOLTAIC ARRAY

Figures 2-16 and 2-17 show several photovoltaic panels arranged in an array. An array can contain any number of modules depending on the output required by the system design and the inverter input range. The panels are 215 watts each with a STC maximum voltage of 42 VDC. There are 10 panels in this array connected in series; therefore, Et = E * En, Et = 420. We can calculate the current of the circuit using Ohm's Law (Figure 2-9), I = P/E where voltage E = 420 VDC and wattage P = 2150 W, then I = 2150 W/420 VDC or I = 5.12 amps.

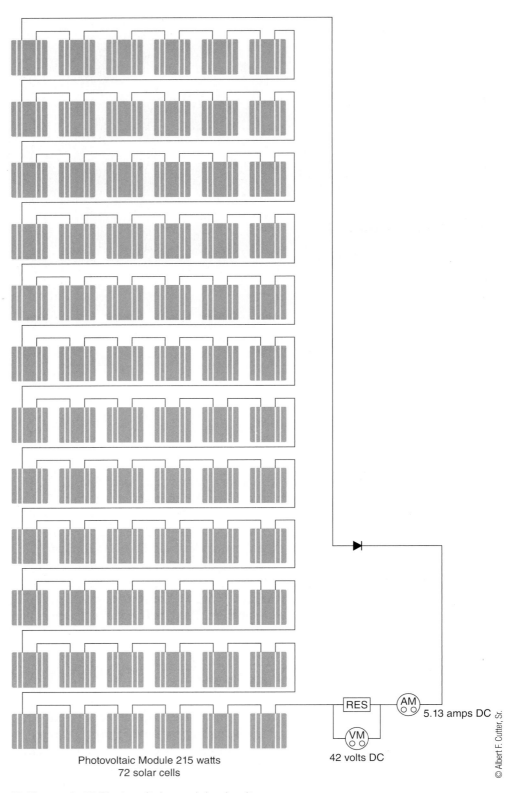

Photovoltaic Module 215 watts
72 solar cells

**Figure 2-15** Photovoltaic module circuit.

■ **Figure 2-16** PV arrays.

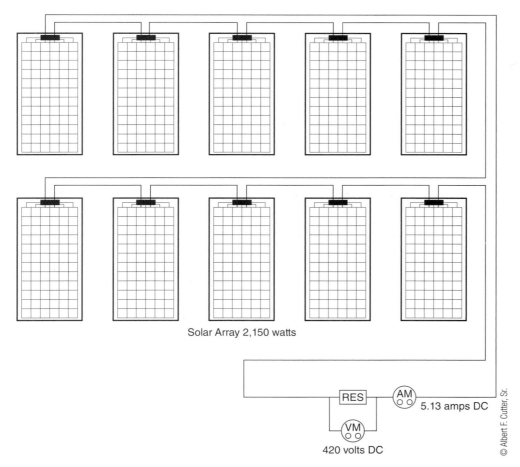

Solar Array 2,150 watts

RES

AM   5.13 amps DC

VM

420 volts DC

■ **Figure 2-17** Photovoltaic string circuit.

## REVIEW

1. The four basic designs of photovoltaic cells are:
   A. Homojunction
   B. Heterojunction
   C. p-i-n/n-i-p
   D. Homogenized
   E. Multijunction

2. What are the three basic formulas of Ohm's Law?

3. What do the following symbols of Ohm's Law represent and what are they measured in?

   **P** =

   **E** =

   **I** =

   **R** =

4. In a series circuit the voltage of the circuit is the sum of the cell voltages.
   A. True
   B. False

5. What does STC stand for?

6. In a parallel circuit the voltage of the circuit is the sum of the cells.
   A. True
   B. False

7. How many panels are in a typical PV array?

8. In a series circuit if the voltage is 120 volts and the current is 10 amps, then what is the resistance?

9. In a parallel circuit if the resistance is 12 ohms and the current is 10 amps, then what is the voltage?

10. According to NEC Article 690.8 the short circuit current is used to calculate the system equipment.
    A. True
    B. False

# CHAPTER 3

# PHOTOVOLTAIC SYSTEM EXAMPLES

## INTRODUCTION

There is much debate about which inverter system design is better. Many will say that one system is more practical than another. Micro-inverter, string inverter, or the central inverter are all good systems and should be used when they meet the requirements of the installation that you are currently working with. Each system's requirements are unique and must be considered on a one-to-one basis. All PV systems are potentially dangerous from an installation and maintenance point of view and require knowledge of electrical systems—both AC and DC systems—and a clear knowledge of the NEC® Code. Systems should only be installed by trained electricians to ensure the safety of all concerned. Every solar panel is an electrical generator; to say that the panels can or should be installed by untrained personnel to save on the system cost is a mistake. One shorted wire can damage an entire system or injure someone. Remember that the best system design can go wrong with one mistake in the installation. The rules in Article 690.4 mandate that only qualified persons install photovoltaic systems. A qualified person is defined in Article 100 as someone that has skills and knowledge in electrical systems installation and has received safety training. (See Article 100 for a complete definition.) When installing any complex electrical system, safety is the first and foremost responsibility of every person involved with the system.

## WHAT WE NEED TO KNOW

Photovoltaic systems are very hard to understand with text only, so in this chapter I will present you with many sample systems. I start off with the simplest of systems and build to more complex systems. It is important to stay with the basics when you look at the systems to follow. If they seem complex, remember that they are just a PV module connected to an inverter. The number of modules and the level of complexity can be confusing; just remember the basics and have fun.

*objective*

The reader will be introduced to the PV systems with the following examples of photovoltaic system installations. We will start with a single-panel, pole-mounted system and then build to more advanced systems. We will explore the systems from an installation and maintenance point of view paying close attention to the National Electrical Code 2011.

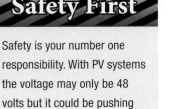

# UTILITY POLE SYSTEM

Figure 3-1 exhibits a single photovoltaic panel with a micro-inverter mounted on a utility pole in New Jersey, United States. There are 200,000 of these systems being installed by a utility company throughout New Jersey. Other states like Florida are installing these systems as well as other installations around the world.

In Figure 3-2 we have a self-monitoring system—a solar panel, inverter, and connection to the electrical grid.

NEC Article 690.2 defines an alternating current (AC) module as a complete self-contained unit with the inverter, solar panel, optics, and other components, exclusive of tracker, designed to generate AC power when exposed to

**Safety First**

Safety is your number one responsibility. With PV systems the voltage may only be 48 volts but it could be pushing 600 amps!!

© Albert F. Cutter, Sr..

■ **Figure 3-1** Pole-mounted AC module.

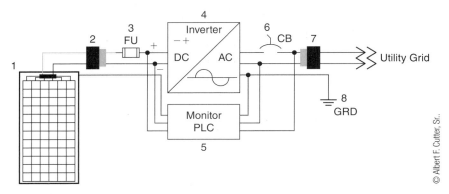

© Albert F. Cutter, Sr..

■ **Figure 3-2** PV AC module.

© Albert F. Cutter, Sr.

■ **Figure 3-3** Inverter on a PV AC module.

light, see Article 690.2 for a complete list of definitions. NEC Article 690.6 sets forth the requirements for an AC module that must be adhered to. The system in Figure 2-3 is an AC-module grid-tied photovoltaic system that uses one panel and a micro-grid-tied inverter, see Chapter 1. This unit uses the power line carrier smart grid technology to monitor the performance of the system. Systems are mounted on light or utility poles and the output of the system is connected directly to the utility grid.

Figure 3-3 shows the installation of the solar panel with the inverter and PLC interface mounted under the solar panel.

With an AC module system we do not calculate the system components, as we would do with a standard system. NEC Article 690.6(A–E) sets forth the requirements of the AC module. See Article 690.6.

(A) **Photovoltaic Source Circuits:** The requirements of Article 690 shall not apply if the connection of the inverter, source circuits, and components are considered as internal wiring of the module.

(B) **Inverter Output Circuit:** The output of the AC module shall be considered an inverter output circuit.

(C) **Disconnect Means:** As required by Articles 690.15 and 690.17, a single disconnecting switch shall be permitted for the combined AC output of one or more AC modules. Each AC module in the system shall be provided with a connector, bolted, terminal-type disconnecting means.

**Safety First**

Knowledge of the NEC code is essential and you must be familiar with the code. I feel that you should not memorize the code; just remember where to look in the code for answers to your questions. The code is written to provide us with a safe system to protect individuals and property. The systems must meet the requirements of the code so familiarize yourself with the code and be safe.

**(D) Ground-Fault Detection:** AC module systems shall be permitted to have a single detection device to detect AC ground faults and disconnect the AC power from the module.

**(E) Overcurrent Protection:** The output circuits of an AC module shall be permitted to have conductor sizing and overcurrent protection in accordance with Article 240.5(b2).

NEC Article 690.52 requires that the AC module be marked with identification of terminals or leads and with the following ratings:

1—Nominal operating AC voltage

2—Nominal operating AC frequency

3—Maximum AC power

4—Maximum AC current

5—Maximum AC module overcurrent protection device rating

The following is a description of Figure 3-2 using the numbered points in the circuit.

1—In the example in Figure 3-2 we have a photovoltaic panel, for the purposes of this discussion we will use the Sanyo 215-watt panel with the specifications in Figures 2-11 and 2-12. This will mean that we have a 215-watt panel with open-circuit voltage of 51.6 volts output at short circuit current of 5.61 amps DC. The maximum circuit voltage should still be calculated for the lowest ambient temperature (a temperature other than 25°C STC).

The voltage temperature coefficient (Vtc) for this panel would be −0.143 V/°C. Using an ambient temperature of 10°C we can calculate the Vmax in the following:

$$Vtc = -0.143 \times (10°C - 25°C)$$

$$Vtc = 2.145$$

$$Vmax = Vtc + Voc$$

$$Vmax = 2.145 + 51.6$$

$$Vmax = 53.745 \text{ VDC}$$

This voltage shall be used to calculate the voltage rating of the cables, disconnects, overcurrent devices, and other equipment that may be used.

The maximum current is the short circuit current of all the modules in parallel; as set forth in Articles 690.8(A)(1) and (B)(1) the multiplication factor will be 156%.

$$I_{max} = Isc * 156\%$$

$$I_{max} = 5.61 * 1.56$$

$$I_{max} = 8.75 \text{ Amps}$$

2—The AC module is not required to have connectors between the solar panel and the inverter but it is usually provided for servicing of the system. If a connector is used, it cannot disconnect the equipment-grounding conductor and it must meet the requirements of NEC Article 690.33.

3—In accordance with NEC Article 240.4 for currents of 800 amps or less the next higher standard fuse can be selected. The fuse would be a 10-amp DC-rated fuse.

4—An inverter is used to convert the DC output of the photovoltaic module to AC voltage required by the grid, see Chapter 1. NEC Article 690.2 defines this system as an interactive system, which requires that it have anti-islanding capability and ground-fault protection, which is usually built into the inverter.

5—The system is not required to have system monitoring. But it is essential for the requirements of the smart grid technology. The system in Figure 3-2 uses a power line carrier (PLC) system. This will monitor critical sensor points like photovoltaic DC output voltage, current, ambient temperature, inverter AC output voltage, and current. This data is transmitted over the power line to a remote monitoring center where the information can be used to monitor the performance of the unit, see Chapter 1.

6—The system must provide a means of disconnecting from the AC power of the utility grid and the requirements of NEC Articles 690.6(C), 690.15, and 690.17 must be met. Also, Article 690.54 requires that the following markings be made at an accessible location at the disconnection means:

    1—Rated output AC current

    2—Nominal operating AC voltage

7—In accordance with Article 690.6(C) the system shall be provided with a connector, bolted, or terminal-type, disconnecting means.

8—Grounding

> This system is an AC module with the grounding taken care of in the assembly of the module. The system has a DC open-circuit voltage of 51.6 volts and, therefore, requires that the DC system be grounded. See NEC Article 690.41, which requires that a DC source system over 50 volts must be grounded, which means that one conductor must be grounded. A smaller panel can be used in the 100-watt range, which would have been under 50 volts. The panel in Figure 2-12 is used since we have the specifications. Also, due to the high efficiency of the SANYO HIT panel, it makes sense to use the highest output panel. NEC Article 690.42 requires that this shall be made at a single point on the photovoltaic output circuit. This is required so that the ground is not supplied from different sources if the ground is broken or removed during maintenance procedures.

> Equipment-Grounding Article 690.43 requires that exposed noncurrent-carrying metals parts of the photovoltaic panels' metal frames, equipment, and conductor enclosures shall be grounded in accordance with Article 250.134 or 250.136(A). Article 690.43(D) requires that if the mounting structure also grounds the frame of the panel, then it must be identified for the purpose. The article continues to describe grounding photovoltaic panel frames to each other. We are working with a single panel in the AC module in this example, so it does not apply.

> Size of Equipment-Grounding Conductors Article 690.45 for the system in the example states that the conductor be sized using the open-circuit current, which is 5.61. For this example, using Table 250.122, the smallest size is #14 AWG. The article further requires that the grounding conductor shall not be smaller than #14 AWG.

> Array Equipment-Grounding Conductor Article 690.46 states that equipment-grounding conductors smaller than #6 AWG shall comply with 250.120(C), which states that a grounding conductor smaller than #6 AWG shall be protected by a raceway or armored cable, or where not subject to physical damage. In the AC module of this example we can say that the ground conductor is protected from damage, as it is part of the integral wiring system of a single-panel AC module.

The Grounding Electrode System is covered in Article 690.47. We are installing an AC Module system so Article 690.47(A) applies to our installation.

The grounding electrode system shall comply with Articles 250.50 through 250.60. Grounding Electrode System Article 250.50 states that all electrodes described in Articles 252(A)(1) through (A)(7) shall be bonded together.

Article 690.47(D), Additional Electrodes for Array Grounding, requires that the grounding electrode shall be installed at the location of ground- and pole-mounted systems.

The grounding conductor electrode shall be no smaller than #14 AWG and must be protected from damage by a raceway or armored cable. The grounding conductor must be continuous—the conductor cannot be spliced or broken. The grounding electrode shall be installed in accordance with Article 250.52 at the location of the pole-mounted system.

# ARRAY

With photovoltaic systems there are a number of panels electrically connected to form arrays. It is important that we understand the different configurations and components of an array, which will be covered in this section.

## Array Definitions

See NEC Article 690.2 for a complete list of PV definitions.

**Array**—An assembly of panels or modules mechanically integrated with a support structure and other components as required—foundation, tracker, etc. That will form a direct-current power-producing unit.

**Monopole Subarray**—A PV array having an output circuit with two conductors— one negative and one positive.

**Subarray**—An "electrical subset" of a photovoltaic array.

**Subset**—"A set that is part of a larger set." Therefore, any discernible electrical portion of an array can be considered a subarray.

**Bipolar Photovoltaic Array**—A PV that has two output conductors. Each of the outputs has an opposite polarity to a center tap or common reference point.

Figure 3-4 shows two monopole arrays combined to form a bipolar array. Arrays are combined to obtain the voltages required by system design and must be done at a combiner box (Figure 3-5). The arrays are connected in parallel circuit at the combiner box shown in Figure 3-6. The combiner box can have disconnects with fuses or finger-safe fuses that disconnect both sides of the circuit and protect the user from a shock.

Article 690.16 Fuses (B) requires that if the fuses cannot be isolated from energized PV circuits when they are serviced, then a disconnect switch must be installed on the photovoltaic output circuit. The disconnect must be within 6 feet (1.8 meters) of the overcurrent devices or a directory showing the location of the disconnect must be installed at the location of the overcurrent devices.

**Safety First**

The arrays are in parallel in the combiner box so care should be taken as both sides of the fuse can be energized.

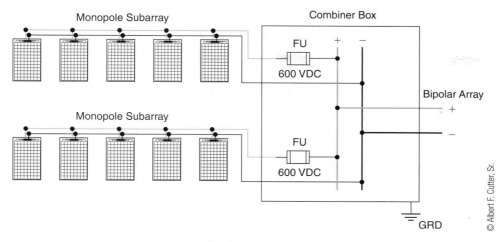

**Figure 3-4** Arrays with a combiner junction box.

**Figure 3-5** SMA combiner box. (See detailed specifications at http://www.sma-america.com.)

Article 690.17 states that the following or equivalent must be marked on the disconnect or adjacent to the disconnect:

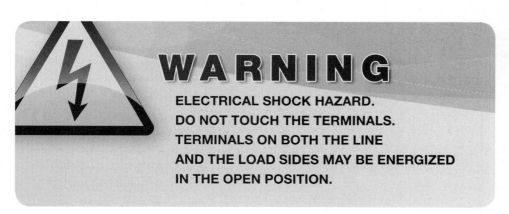

**WARNING**

ELECTRICAL SHOCK HAZARD.
DO NOT TOUCH THE TERMINALS.
TERMINALS ON BOTH THE LINE
AND THE LOAD SIDES MAY BE ENERGIZED
IN THE OPEN POSITION.

**Safety First**

Most often PV output circuits are connected in parallel at the inverter. Caution must be taken as the line side of the overcurrent device may be energized.

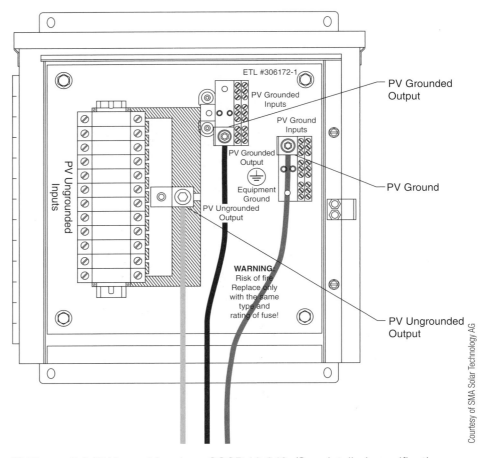

Courtesy of SMA Solar Technology AG

■ **Figure 3-6** SMA combiner box, SCCB 12-240. (See detailed specifications at http://www.sma-america.com.)

The DC open voltage and current can be very high in the array. Caution must be taken to avoid a fatal shock or a high voltage arch, which can cause serious injury or loss of life. Remove the fuses or open the disconnect switches then connect all the circuit conductors, terminating them in the combiner box. Connect the solar panels last. Remember that safety is job one, and it is the responsibility of each person for their own safety and the safety of everyone around them.

In a parallel circuit the current is additive array current total $I_t = I_1 + I_2$ or $I_t = I_n * I$. In this example, the current $I = 5.13$, which is $I_t = 2 * 5.13$; therefore, $I_t = 10.26$ at a voltage of $E_t = 210$ VDC. The wattage is $P_t = E_t * I_t$, then $P_t = 210 * 10.24$, then $P_t = 2,150$ watts or 2.15 kW.

## Identification and Grouping

Article 690.4(B) 1 through 4 sets forth the rules for the grouping and identification of photovoltaic circuits. It is very important that circuits from other non-PV circuits not be combined with PV circuits in the same raceway, cable, outlet box, cable tray, or junction box. The parent text of Article 690.4(B) makes it very clear that if they are combined they must be separated by a partition. But it makes no sense to combine the circuits for other non-PV systems with the PV system conductors. This could cause problems in the future when the system needs maintenance or repair. The text continues to state that the conductors are to be grouped and identified as required by Articles 690.4(B)(1–4) and that the conductors shall be marked by separate color coding, tagging, marking tape, or other approved means, as shown in Figure 3-7.

1. Requires that photovoltaic source circuits be identified at all points of termination splices and connections. This would include all junction boxes and combiner boxes.

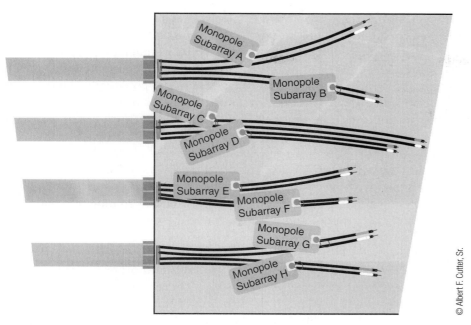

© Albert F. Cutter, Sr.

**Figure 3-7** Grouping and identification of PV circuit conductors.

2.  Mandates that photovoltaic output and inverter circuits be identified at all points where they are spliced, terminated, and connected. This would include the combiner boxes, disconnects, and the inverter connections, as well as at all splice points.

3.  Mandates that when conductors of more than one PV system occupy the same raceway, junction box, or equipment, the conductors of each system must be clearly marked and identified at all splices and termination points. There is an exception to this requirement, which is when the indemnification of the conductors is evident by the arrangement or spacing of the conductors.

4.  Sets forth the requirements of grouping conductors in a junction box or raceway with removable covers. The AC and DC conductors of each system shall be grouped at least once with wire ties or similar means and then again at intervals not to exceed 6 feet (1.8 meters). There is an exception that if the circuit enters from a raceway or cable that makes the grouping obvious, then the grouping is not necessary.

## Qualified Persons

Article 690.4(E) mandates that only qualified persons shall install photovoltaic systems. It is commonly understood that systems covered under the electrical code only be installed and maintained by qualified personnel that are well trained in electrical systems and the safety requirements. Due to the proliferation of PV systems and the allure of a growing market, less-than-qualified entrepreneurs are trying to get into the business. As a result there are plumbers, landscapers, roofers, and handymen getting into the solar business with little or no electrical

and safety training. These are electrical systems and should only be installed by qualified electricians and contractors. Many states and local governments are passing laws that require that a PV system must be installed by trained personnel and some go further to require specific credentials for contractors that install PV systems. But it is more than that a PV system that is potentially dangerous at a worksite and, as with any electrical systems, if installed incorrectly, the system will not perform properly and in the worst case it could cause a fire or electrical shock.

Article 640.4(F) sets forth the rules for the routing of photovoltaic source and output conductors. This is to protect the firefighters from cutting into a live PV circuit when they are venting a roof. This requires that the PV circuits be physically routed along building structural members. This means routing the conductor along beams, trusses, rafters, and columns where these members can be located through observation.

This requirement also requires that if the PV system is embedded in the roof in areas that are not covered by PV modules and other equipment, the location of the circuits shall be clearly marked. There is no description of how the locations should be marked. If the location of the system is where there is snow, then the markings should be elevated so that they can be seen after a snow fall.

## Bipolar Photovoltaic Systems

There is a potential danger of combining the circuits of two monopole subarrays where the sum of the voltages, without consideration of the polarity, will exceed the rating of the conductor insulation and voltage rating of equipment in a bipolar array and that all conductors of a subarray must be physically separated. Article 690.4(G) recognizes this potential and sets forth that all the conductors from each monopole subarray be routed in the same raceway until connected at the inverter. It also sets forth that the overcurrent and disconnecting means for each monopole subarray output must be in separate enclosures. There is an exception that allows the overcurrent protective devices and disconnecting means for each monopole subarray to be installed in the same switchgear provided that it is listed switchgear rated for the maximum voltage between the circuits and has a physical barrier that separates the disconnecting means from each monopole subarray.

## Arc-Fault Circuit Protection (Direct Current)

Photovoltaic systems are exposed to extreme temperatures, wind, snow, rain, and dirt and over time this can cause deterioration and failure of system components. This can cause a DC arc fault that can easily cause a fire on the structure or damage to system components. Article 690.11 sets forth the requirement that a system with a maximum PBV DC voltage of 80 volts or greater must have DC arc-fault circuit protection device (AFCP). These devices must detect and interrupt the DC arc faults and then disconnect or disable the inverters or charge controllers and all the system components in the faulted circuit. The AFCP must have a visual annunciation that must be reset manually to insure that the fault was observed by personnel.

# STRING INVERTER SYSTEM

In Figure 3-8 you can clearly see the string of Sunny Boy inverters. In the background under the array you can see the SMA combiner boxes.

Figure 3-9 shows a system with the inverters mounted remotely inside the building. This configuration has additional requirements to meet the code. We will cover this further in the system review.

Courtesy of SMA Solar Technology AG

■ **Figure 3-8** String inverter system. (See detailed specifications at http://www. sma-america.com.)

Courtesy of SMA Solar Technology AG

■ **Figure 3-9** String array with SMA Sunny Boy inverters. (See detailed specifications at http://www.sma-america.com.)

■ **Figure 3-10** Sanyo solar arrays. (See us.sanyo.com for complete specifications.)

Figure 3-10 shows a large rooftop array of the Sanyo HIT panels.

Figure 3-11 is a photovoltaic system using a string inverter configuration with an output of 17.2 KW STC. STC values are used as it is the only way to get an output that we can compare to other equipment configurations. But we must use the open circuit voltage and short circuit current for calculation of the system circuit components.

In a string inverter system the inverter should be as close to the array as possible, as shown in Figure 3-8, to reduce DC losses due to voltage drop. For example, if the array is 200 feet from the input of the inverter and is carrying 8 amps at 420 VDC and the wire size is #14 AWG. $V_d = 2 \times K \times L \times I/CMIL$, we know from Chapter 1 that K = 12 for copper wire, L is the one-way length of the cable run 200 feet, and I = short circuit current 5.61. The code requires that we multiple this by 125%. According to Article 690.8(A) the current would be 7.01. That seems to be enough but Article 690.8(B) requires that the current is continuous and must be factored by another 125%. So the correct factored current will be 8.77 amps. The voltage drop $V_d$ is calculated as $V_d = 2 \times 12 \times 200 \times 8.77/4110$.

$$V_d = 10.24$$

$$V_d\% = V_d/V * 100$$

$$V_d\% = 10.24/420$$

$$V_d\% = 2.43\%$$

The voltage drop is calculated at 2.43%, close to the NEC requirement that the total voltage of the circuit drop must be 3% or less. We must remember that the 3% rule is for the entire wiring system not just the DC photovoltaic source circuits. With the high cost of kW we do not want to lose any kW to heat losses. Locating

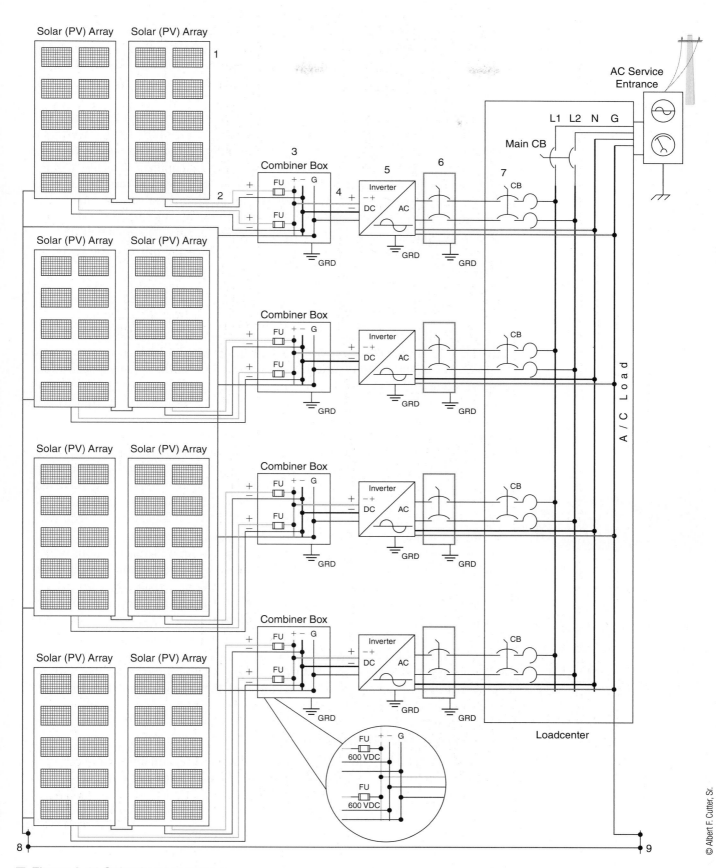

■ **Figure 3-11** String inverter system.

the combiner boxes closer to the photovoltaic cell would reduce this. If we move the combiner boxes 100 feet closer, the $V_d$% would be 1.2%; or, we could change the wire size to #12 AWG and the 200 foot $V_d$% would be 1.5%. You can see the location of the system components and the wire length and size of the conductor is very important.

Arranging the inverters in the string configuration allows for a distributed system with shorter DC cable runs. In a string system if one of the string inverters is out for service, the other inverter(s) will stay online producing an output. Also the string inverter system allows for monitoring at more data points. Using SMA Sunny Boy inverters, which can be monitored from a central location over an Ethernet network, the performance of smaller arrays can be monitored, see Chapter 1. It is very important that the system be monitored with the most number of data points so that maintenance and repairs can be made quickly and efficiently.

Using the system in Figure 3-11 we will look at points that must be considered.

## 1 – Photovoltaic Array

Solar arrays are made of solar modules that are assembled into panels that are connected in series or parallel as required; as shown is Figure 2-17 this array is connected in series. The configuration of the arrays is determined by the system specifications and the requirements of the code. The arrays must be sized and configured to meet the specified voltage and current to match the inverter specifications—in this case 600 volts maximum (see Appendix). It is common practice to talk about the arrays output in terms of maximum STC voltage and current. People will refer to arrays as 42-volt DC panels connected in series to produce 420-volt DC arrays with an output current of 5.13 amps—these are the STC values. To install and maintain the arrays we must use the open-circuit voltage of 51.6-volt DC and the short circuit current of 5.61-amp DC as required by the NEC 2011. This means that the unadjusted open-circuit voltage of the array is 516 volts DC. NEC Article 690.7.C requires a maximum system voltage of 600 volts for a residential solar system.

We will use the jumpers that are supplied by the manufacturer to connect the panels together; this will comply with Article 690.33. The connectors must be locking or latching type and systems operating over 30 V must require use of a tool to open if they are in locations that are accessible. This will allow removal of a single panel for maintenance or replacement. These are #12 AWG wire with locking type MC4™ connector; see the specifications shown in Figure 2-13. See Appendix Figures A-1, A-2, and A-3 for details on the connectors. Article 690.33 states that connectors require a polarized connector, which are not interchangeable with receptacles used in different electrical systems, and they shall be a locking connector as stated in sections (A)–(E) of the article.

### Installation and Maintenance

NEC 690.18 states that short-circuiting, open-circuiting, and an opaque covering shall be used to disable the array or any portions of the array.

### Wiring Methods at the Array

Article 690.31 allows the use of single-conductor type USE-2 and single conductor cable labeled and listed for photovoltaic to be used in exposed outdoor locations in the photovoltaic source circuits for photovoltaic panel inter-connection within the array.

**Safety First**

Caution is needed when working with photovoltaic modules. When exposed to sunlight the modules will be energized and could expose the personnel to an electric shock or DC arcing.

### Access to Boxes

Boxes under the panels in Figure 3-11 must be accessible. Article 690.34 is the rule that covers junction, pull, and outlet boxes. If the boxes are not directly accessible, boxes are considered accessible if the module or panel over them is secured with removable fasteners and the flexible wiring method is used.

### Ungrounded PV Power Systems

Photovoltaic systems are allowed to be ungrounded provided that they comply with Articles 690.35(A)-(G). The system in Figure 3-11 is a grounded system.

### System Grounding

The system in Figure 3-11 is over 50 volts. According to Article 69.41 the system must have a grounded circuit conductor. According to Article 690.42 the DC grounding conductor must occur at only one point. There is an exception that requires that the connection correlates with the operation of the GFP (see Article 690.5), which in Figure 3-11 is at the inverter. Now we must deal with the connection of the grounding electrode conductor, the equipment-grounding conductor, and the ground DC conductor. This connection is covered in Article 250.168 where an unspliced bonding jumper must be installed at this point. This jumper must be sized the same way as the grounding electrode conductor for the system. According to Article 250.166(B) the grounding electrode conductor shall not be smaller than the largest DC circuit conductor supplied by the system, and not smaller than #8 AWG copper or #6 AWG aluminum.

### Equipment Grounding

Article 690.43 requires that all noncurrent-carrying metal parts of the PV array and conductor enclosures and support hardware shall be grounded in accordance with Article 250.134 or 250.136(A) regardless of voltage of the system. Devices that are listed and identified for grounding and/or bonding the metallic frames of PV modules shall be permitted to bond the exposed metallic frames of PV modules to adjacent PV modules and grounded mounting structures. The panel frames are usually aluminum and care must be taken to properly make a grounding connection. A copper-bodied lug with a tin coating that is listed for direct burial must be used. These lugs are compatible with aluminum surfaces and the stainless steel hardware and will survive the elements in an outdoor environment. These lugs should be fastened with a machine screw, not a sheet metal screw, as this will better secure them. Listed modules will usually have threaded inserts in the frame for the grounding screws and should be used, see Figure 2-12. The screw and the mounting surface of the lug must be coated with an antioxidant compound rated for aluminum connections. This ensures that aluminum oxide will not form and provide a low-impedance long-lasting connection.

In accordance with Article 250.110, an equipment-grounding conductor is required between the PV array and other equipment. This means that the grounding conductor must be carried with the DC conductors and ground the enclosures, boxes, and raceways. This is further strengthened by the statement in the code that equipment-grounding conductors for the PV array shall be contained within the same raceway or cable, or otherwise run with the PV array circuit conductors when those circuit conductors leave the vicinity of the PV array.

### Size of Equipment-Grounding Conductors

The sizing of the equipment-grounding conductors must be in accordance with Article 690.45, which states that it must follow Table 250.122. If no overcurrent protection device is used in the circuit, then the size is based on the short circuit of the PV. In the system in Figure 3-11 the overcurrent device in the combiner box protects the photovoltaic source circuit and it is rated at 10 amps. If no overcurrent protective device is used in the circuit, then the photovoltaic rated short-circuit current shall be used in Table 250.122. It is not required to increase the size of the equipment-grounding conductor for voltage drop. The equipment-grounding conductors shall be no smaller than #14 AWG. The equipment-grounding conductor in the system in Figure 3-11 would be #14 AWG copper.

### Array Equipment-Grounding Conductors

Article 690.46 covers the array equipment-grounding conductor. As stated, if the conductor is smaller than #6 AWG, it must comply with Article 250.120(C). This requires that grounding conductors smaller than #6 AWG must be protected from damage by raceway or armor at the array.

### Grounding Electrode System

Article 690.47 covers the requirements for the system grounding electrode(s). AC and DC systems follow the Article 250 requirements for grounding systems. The system in Figure 3-11 must comply with Article 690.47(C)(1) for systems with AC and DC grounding requirements.

> 1—Requires that the AC and DC grounding system be bonded together. This is done in Figure 3-11 at points 8 and 9.

> 2—The bonding conductor size is based on the largest of the size of the AC grounding conductor or DC grounding conductor based on Article 250.166(C), which states that the conductor size shall not be required to be larger than #6 AWG copper wire or #4 AWG aluminum wire.

### Continuity of Equipment-Grounding Systems

Article 690.48 requires that if equipment is removed for service or repair, the system ground jumper must be installed to maintain continuity of the system ground. This requires that if the combiner box or inverter is removed, then a jumper must be installed while the equipment is removed.

## 2 – Photovoltaic Source Circuit

Article 690.2 defines the photovoltaic source circuit as the circuit from the photovoltaic panel to a common DC junction point and between panels. The size of the circuit conductors must be done in accordance with Articles 690.8(A-1) and (B-1). In accordance with A-1, the sum of the short circuit current of the parallel-connected modules must be multiplied by 125%. In Figure 2-13 the short circuit current is 5.61 A, therefore, the short current (Isc) is multiplied by 1.25. Isc * 1.25 so the following is 5.61 A * 1.25 = 7.01 A. But the code goes further in B-1, which states that the overcurrent devices and the conductors are sized to carry not less than 125% of the maximum currents as calculated in Article 690.8(A). Therefore the 7.01 calculated in A-1 is multiplied by 125% (7.01 A * 1.25 = 8.76 A).

## Conductors Size

To determine the correct wiring size for the photovoltaic source circuit, we must start with the current and voltage. In Figure 2-25 the corrected current is 9 A and the voltage at the photovoltaic source DC voltage is 516 volts DC. Then the ambient temperature, conduct fill, and the voltage drop must be factored.

A factor that must be dealt with is ambient temperature when the conductors are run with the array temperatures 70°C or higher. This means that the wiring, and any nonmetallic raceway if used, must have a minimum of 90°C rating. For example, the 9-amp load in Figure 3-11 the #14 AWG THWN-2 has a current capacity of 25 A on Table 310.16 for 70°C. After applying a de-rating factor of 0.58 (from Article 310.16), this would be 14.5 A, which will work for the 9-amp load.

Voltage drop should always be reviewed at this point. For example, if the array is 200 feet from the PV to the combiner box and is carrying 9 amps at 516 VDC and the wire size is #14 AWG. $V_d = 2 \times K \times L \times I/CMIL$, we know from Chapter 1 that K = 12 for copper wire, L is the one-way length of the cable run (200 feet), and I = adjusted short circuit current 9 A. The voltage drop $V_d$ is calculated as:

$$V_d = \frac{2 \times 12 \times 200 \times 9}{4110}$$

$$V_d = \frac{43,200}{4110}$$

$$V_d = 10.51 \text{ volts}$$

$$V_d\% = \frac{V_d}{V}$$

$$V_d\% = \frac{10.51}{516}$$

$$V_d\% = 2.04\%$$

This would be too high for this segment of the circuit; it would be better to increase the size to #12 AWG, the savings in kWs will more than pay for the increase in wire costs.

## Wiring Methods

The wiring methods must meet the requirements of Article 690.31(A). This rule of the code allows for the use of all raceways and cable wiring methods in the code, and those other wiring systems and fittings that are listed and labeled for use in photovoltaic arrays shall be permitted.

Article 690.43 requires that the equipment-grounding conductors for the photovoltaic array shall run with the photovoltaic circuit either contained in the raceway or cable, or run with the conductors when they leave the vicinity of the array.

Article 690.31(A) states that where photovoltaic source and output circuits that have maximum operating system voltages greater than 30 volts and are installed in readily accessible locations, circuit conductors shall be installed in a raceway.

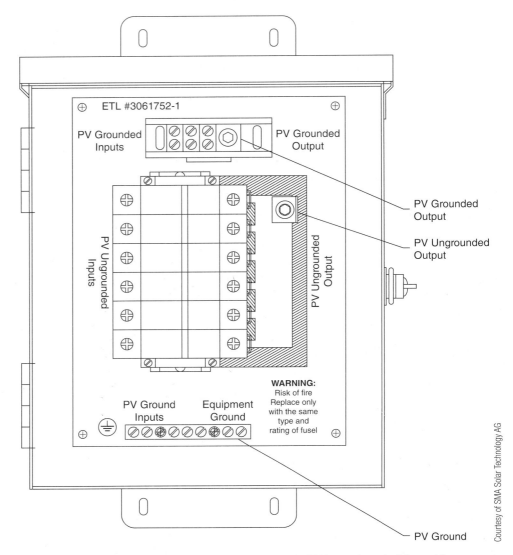

■ **Figure 3-12** Combiner box wiring, SBCB-6-90 6 PV input fuse holders, 15 amp maximum fuse size. (See detailed specifications at http://www.sma-america.com.)

## 3 – Combiner Box

The combiner box (Figure 3-12) allows for the convenient termination of the photovoltaic source circuits. It contains the fuses for the photovoltaic circuit and provides the disconnect means for the photovoltaic circuits. See Figure 3-11.

### Disconnecting Requirements

NEC 690.13–18 sets forth the requirements for the disconnecting means that are required for photovoltaic systems.

NEC 690.13 requires that there shall be a means to disconnect all current-carrying conductors of the photovoltaic circuit from all other conductors in a building or other structure. It also states that a switch, circuit breaker, or other disconnect device—either AC or DC—be installed in the ungrounded conductor if the operation of the device leaves the marked, grounded conductor in an ungrounded and energized state. The grounded conductor may have a terminal or bolted connector that allows qualified personnel to perform maintenance or troubleshooting of the equipment. This is very important as this can cause physical injury and damage to the photovoltaic equipment.

There is an exception to the requirements of Article 690.13. If the switch or circuit breaker is part of the ground fault detection system required by Article 690.5, it shall be permitted to open the grounded conductor of the photovoltaic circuit when the device is automatically opened as a normal function of the device in responding to ground faults. The device shall indicate the presence of a ground fault.

### Disconnect Marking

NEC 690.14(C-2) requires that each disconnecting means in a photovoltaic system shall be permanently marked to identify it as a photovoltaic disconnect.

NEC 690.17 requires that where the disconnecting means may be energized in the open position, a sign shall be mounted on or adjacent to the disconnecting means. The sign shall be clearly legible and have the following words or equivalent:

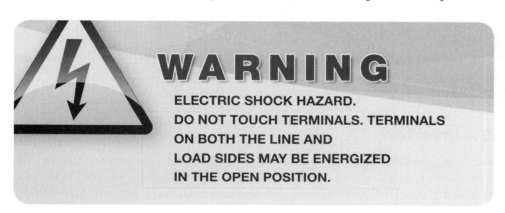

**WARNING**

ELECTRIC SHOCK HAZARD.
DO NOT TOUCH TERMINALS. TERMINALS
ON BOTH THE LINE AND
LOAD SIDES MAY BE ENERGIZED
IN THE OPEN POSITION.

### Maximum Number of Disconnects

NEC 690.14(C-4) states that the maximum number of disconnecting means in a photovoltaic system should not exceed six switches or circuit breakers. In a string array each array inverter combination is a separate photovoltaic system. Therefore the system in Figure 3-11 is correct. This does not mean that you cannot have more than six switches, fuses, or circuit breakers in a system, it means that you must provide a single disconnecting means in the circuit to disconnect a multi-fuse combiner box. This is not required in the system in Figure 3-11.

### Fuses

Fuses in the combiner box are energized from both sides when the circuit is energized. This is due to that fact that the photovoltaic source circuits are connected in parallel and the fuses are fed from the buss and the circuit connected. Article 690.16 requires that a disconnecting means be provided to disconnect a fuse from all sources of the supply if the fuses are accessible to other than qualified persons. Such a fuse in the photovoltaic source circuit shall be disconnected independently of fuses in other circuits. In the SMA combiner boxes shown in Figure 3-11 there are finger safe fuses that disconnect the circuit without allowing the personnel to touch the fuse/fuse holder. Also there is a separate fuse for each photovoltaic circuit.

### Fuse Sizing

In an electrical system, fuses are used to protect the wiring and equipment from excessive currents. If these currents are allowed, they will cause damage, heating, or in the extreme can cause fire. If the fuse is too small, it could open during normal operation. If the fuse is too large, it cannot provide the required protection.

**Safety First**

Care must be taken when working with photovoltaic source circuits. Never disconnect the ungrounded circuit under load and never remove the fuse under load. This will cause DC arcing, which can cause physical injury and damage the fuse holder and circuit buss. It is best to cover the array and disconnect the circuit from the load whenever working on the system.

**Safety First**

It is important for the safety of others that you clearly mark the disconnect means. As the installer you are aware of the safety hazard. When the system is serviced sometime in the future, it is important to warn the servicing personnel.

The minimum sizes of the fuses are calculated using the short circuit current (Isc) of the photovoltaic source circuit. The NEC 690.8(A and B) requires that all fuses be sized for a minimum of 1.56 times the Isc of the circuit. In the photovoltaic source circuit in Figure 3-11 the Isc = 5.61Adc, then we would determine the fuse size by Fs = 5.61 * 1.56, Fs = 8.75. The next standard fuse size would be a 10 A, 600 VDC fuse.

## 4 – Photovoltaic Output Circuit

### Wire Size

The photovoltaic source circuits in the combiner box are connected in parallel so the short circuit (Isc) is the sum of the Isc of the individual circuits. In the system in Figure 3-11, the Isc = Isc1 + Isc2, then Isc = 5.61 + 5.61 or Isc = 11.22 Adc. As required by 690.8(A and B) the Isc must be multiplied by 1.56, then Isc = 11.22 * 1.56 or Isc = 17.50 Adc. The wire size must be determined using Table 310.16. The wire size must be adjusted in the field for conditions, such as the mounting location of the inverter. If the conductors are exposed to sunlight, then they must be de-rated. Also, conduit fill and voltage drop must be considered.

## 5 – Inverter

For complete details on the inverter refer to Chapter 1. The inverter used in the example in Figure 3-11 has been designed with SMA inverters, which provides the required ground fault protection (GFP) and anti-islanding.

In a string array system there is a number of grid-tied inverters in or on the building of the installation. Article 690.4(H) clearly permits this type of installation. This requirement mandates that if the inverters are remote from each other, a directory in accordance with Article 705.10 shall be installed at each DC photovoltaic disconnecting means, at each of the AC disconnection means, and at the main service disconnection means. This directory must show the location of all AC and DC PV system disconnecting means in or on the building. There is an exception to this article where all the DC PV disconnection means and inverters are grouped at the main service disconnection means the directory is not required.

### Ground Jumper

NEC 690.49 requires that if the inverter in an interactive utility connected system is removed for maintenance, a bonding jumper shall be installed to maintain the system grounding while the inverter is removed. Bonding jumpers must be done in accordance with Article 250.120(C).

### Wire Size

NEC 690.8 requires that the maximum continuous output current of the inverter shall be used. The specification of the inverter must be consulted for the value. In the system in Figure 3-11 the maximum current (Im) is 29 Aac (see specifications in Appendix Figure A-5). As required by Article 690.8(A-B), the Im must be multiplied by 1.56, then Im = 29 * 1.56 or Im = 45.24 Aac. The wire size must be determined using Table 310.16. The wire size must be adjusted in the field for conditions, such as the mounting location of the inverter. If the conductors are exposed to sunlight, then they must be de-rated. Also, conduit fill and voltage drop must be considered.

## 6 – Disconnect Means

NEC 690.14(C) states that means shall be provided to disconnect all conductors in a building or other structure from the photovoltaic system conductors.

In accordance with NEC 690.15, means shall be provided to disconnect the inverter(s) from all ungrounded conductors of all sources.

NEC 690.17 requires that where the disconnecting means may be energized in the open position, a sign shall be mounted on or adjacent to the disconnecting means. The sign shall be clearly legible and have the following words or equivalent:

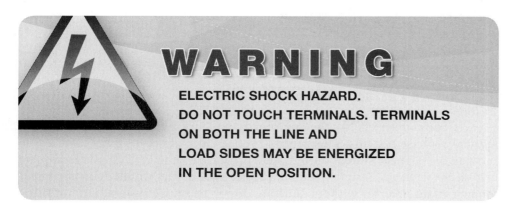

## 7 – Circuit breaker

The circuit breaker is the point of connection of the interactive inverter system. The rules of Article 690.45 must be followed for the connection to the main panel. In the system in Figure 3-11 the connection is made to the load side of the service entrance equipment and must follow the rules in Article 690.45(B1) through (B7).

> 1—It is required that each inverter or interconnecting system must have a dedicated fuse or circuit breaker.
>
> 2—The sum of the current rating of the overcurrent devices shall not exceed 120% of the rating of the buss bar or conductor.
>
> 3—The connection point shall be on the line side of the GPF equipment.
>
> 4—Equipment containing multiple devices with multiple sources shall be marked to indicate the presence of all sources.
>
> 5—If the circuit breakers are back fed, then the circuit breaker shall be suitable for such operation. If a circuit breaker is marked line and load, then it has been tested only in that direction.
>
> 6—Listed plug-in-type back fed circuit breakers from utility-interactive inverters complying with Article 690.60 shall be permitted not to have the additional fastener normally required by Article 480.36(D). This is allowed because if the breakers are lifted from the bus bar, then the inverter would trigger the anti-islanding function and shut off the AC output immediately as per Article 690.61.
>
> 7—The circuit breakers feeding the panel must be at the opposite end of the bus bar (load end) from the main breaker (line end) and be permanently marked with the following or equivalent.

**WARNING**

INVERTER OUTPUT CONNECTION
DO NOT RELOCATE THIS OVERCURRENT
DEVICE

# CENTRAL INVERTER

Central inverters like the unit shown in Figure 3-13 are used in large commercial 3-phase systems. This central inverter is made up of three inverters to provide the 3-phase system output that is required. In many systems redundant central inverters are installed with a transfer system that would allow for maintenance of an inverter without shutting down the entire system. Do not think of a central inverter system as only very large systems. Most small residential systems are central inverter systems. The system that I have laid out in Figure 3-14 shows a commercial system for the purpose of reviewing the system and NEC requirements. A central inverter system is defined simply as a PV system with a single inverter system.

Figure 3-14 shows a design for a centralized system; this system has multiple arrays feeding a central inverter.

Using the system in Figure 3-14 we will look at points that must be considered.

Courtesy of SMA Solar Technology AG

■ **Figure 3-13** SMA central inverter system. (See detailed specifications at http://www.sma-america.com.)

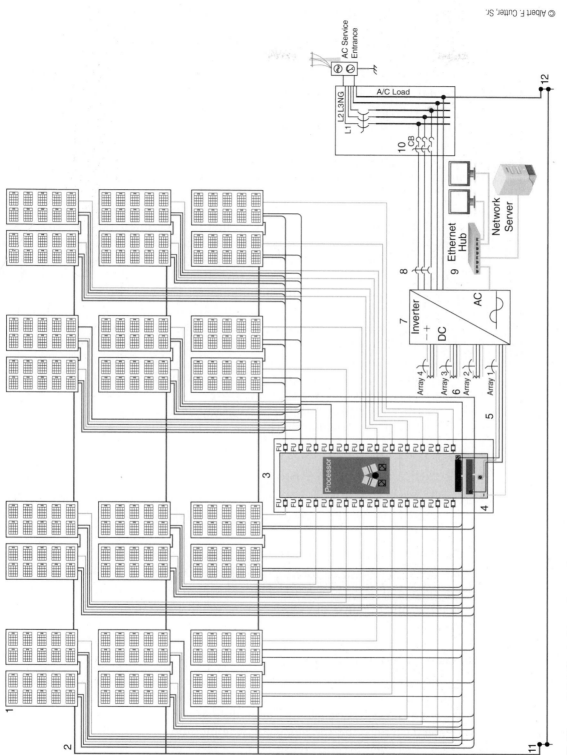

**Figure 3-14** Central inverter system.

## 1 – Photovoltaic Array

PV system arrays are made of solar modules that are assembled into panels as shown is Figure 2-15; these modules are connected in series and the strings are connected in parallel. A string is defined by a group of panels connected in series to provide the necessary outputs for the system. The configuration of the arrays is determined by the system specifications and the requirements of the code. The arrays must be sized and configured to meet the specified voltage and current and must be configured for the inverter specifications, in this case 600 volts maximum (see Appendix). It is common practice to talk about the arrays output in terms of maximum STC voltage and current. We will refer to an array as 42-volt DC panels connected in series to produce 420-volt DC arrays with an output current of 5.13 amps—these are the STC values. To install and maintain the arrays we must use the open-circuit voltage of 51.6-volt DC and the short circuit current of 5.61-amp DC as required by the NEC 2011. This means that the unadjusted open-circuit voltage of the array in Figure 3-14 is 516 volts DC. Connecting the panels together we will use the jumpers that are supplied by the manufacturer; this will comply with Article 690.33. The connectors must be locking or latching type and on systems operating over 30 V must require use of a tool to open if they are in locations that are accessible. This will allow removal of a single panel for maintenance or replacement. The jumpers are #12 AWG wire with locking type MC4™ connector; see the specifications shown in Figure 2-13. See Appendix Figures A-1, A-2, and A-3 for details on the connectors. Article 690.33 states that connectors require a polarized connector, which are not interchangeable with receptacles used in different electrical systems, locking connector in sections (A)–(E) of the article.

**Safety First**

Caution is needed when working with photovoltaic modules. When exposed to sunlight the modules will be energized and could expose the personnel to an electric shock or DC arcing.

### Installation and Maintenance

NEC 690.18 states that short-circuiting, open-circuiting and an opaque covering shall be used to disable the array or any portions of the array.

### Wiring Methods at the Array

Article 690.31 allows the use of single-conductor type USE-2 and single conductor cable labeled and listed for photovoltaic to be used in exposed outdoor locations in the photovoltaic source circuits for photovoltaic panel inter-connection within the array.

### Access to Boxes

Boxes under the panels in Figure 3-14 must be accessible; Article 690.34 is the rule that covers junction, pull, and outlet boxes. If the boxes are not directly accessible, boxes are considered accessible if the module or panel over them is secured with removable fasteners and the flexible wiring method is used. This is the method used with the panels used in Figure 3-14.

### Ungrounded PV Power Systems

Photovoltaic systems are allowed to be ungrounded provided that they comply with Articles 690.35(A)–(G). The system in Figure 3-14 is a grounded system.

## System Grounding

The system in Figure 3-14 is over 50 volts according to Article 69.41; the system must have a grounded circuit conductor. According to 690.42 the DC grounding conductor must occur at only one point. There is an exception that requires that the connection correlates with the operation of the GFP, see Article 690.5, which in Figure 3-14 is at the inverter. Now we must deal with the connection of the grounding electrode conductor, the equipment-grounding conductor, and the ground DC conductor. This connection is covered in Article 250.168 where an unspliced bonding jumper must be installed at this point. This jumper must be sized the same way as the grounding electrode conductor for the system. According to 250.166(B) the grounding electrode conductor shall not be smaller than the largest DC circuit conductor supplied by the system, and not smaller than #8 AWG copper or #6 AWG aluminum.

## Equipment Grounding

Article 690.43 requires that all noncurrent-carrying metal parts of the PV array and conductor enclosures and support hardware shall be grounded in accordance with Article 250.134 or 250.136(A) regardless of voltage of the system. Devices that are listed and identified for grounding and/or bonding the metallic frames of PV modules shall be permitted to bond the exposed metallic frames of PV modules to adjacent PV modules and grounded mounting structures. The panel frames are usually aluminum and care must be taken to properly make a grounding connection. A copper-bodied lug with a tin coating that is listed for direct burial must be used. These lugs are compatible with aluminum surfaces and the stainless steel hardware and will survive the elements in an outdoor environment. These lugs should be fastened with a machine screw, not a sheet metal screw, as this will better secure them. Listed modules will usually have threaded inserts in the frame for the grounding screws and should be used, see Figure 2-12. The screw and the mounting surface of the lug must be coated with an antioxidant compound rated for aluminum connections. This ensures that aluminum oxide will not form and provide a low-impedance long-lasting connection.

In accordance with Article 250.110, an equipment-grounding conductor is required between the PV array and other equipment. This means that the grounding conductor must be carried with the DC conductors and ground the enclosures, boxes, and raceways. This is further strengthened by the statement in the code that equipment-grounding conductors for the PV array shall be contained within the same raceway or cable, or otherwise run with the PV array circuit conductors when those circuit conductors leave the vicinity of the PV array.

## Size of Equipment-Grounding Conductors

The sizing of the equipment-grounding conductors must be in accordance with Article 690.45, which states that it must follow Table 250.122. If no overcurrent protection device is used in the circuit, then the size is based on the short circuit of the PV. In the system in Figure 3-14 the overcurrent device in the combiner box protects the photovoltaic source circuit and it is rated at 10 amps. If no overcurrent protective device is used in the circuit, then the photovoltaic rated short-circuit current shall be used in Table 250.122.

It is not required to increase the size of the equipment-grounding conductor for voltage drop. The equipment-grounding conductors shall be no smaller than #14 AWG. The equipment-grounding conductor in the system in Figure 3-14 would be #14 AWG copper.

### Array Equipment-Grounding Conductors

Article 690.46 covers the array equipment-grounding conductor. As stated, if the conductor is smaller than #6 AWG, it must comply with Article 250.120(C). This requires that grounding conductors smaller than #6 AWG must be protected from damage by raceway or armor at the array.

### Grounding Electrode System

Article 690.47 covers the requirements for the system grounding electrode(s). AC and DC systems follow the Article 250 requirements for grounding systems. The system in Figure 2-25 must comply with Article 690.47(C)(1) for systems with AC and DC grounding requirements.

> 1—Requires that the AC and DC grounding system be bonded together. This is done in Figure 3-14 at points 8 and 9.

> 2—The bonding conductor size is based on the largest of the size of the AC grounding conductor or DC grounding conductor based on Article 250.166(C), which states that the conductor size shall not be required to be larger than #6 AWG copper wire or #4 AWG aluminum wire.

### Continuity of Equipment-Grounding Systems

Article 690.48 requires that if equipment is removed for service or repair, the system ground jumper must be installed to maintain continuity of the system ground. This requires that if the combiner box or inverter is removed, then a jumper must be installed while the equipment is removed.

## 2 – Photovoltaic Source Circuit

Article 690.2 defines the photovoltaic source circuit as the circuits from the photovoltaic panel to a common DC junction point and between panels. In Figure 3-14 there are 24 strings; each string has 10 series connected panels. The string open-circuit voltage is 516 volts DC.

$$E = E * n$$

$$E = 51.6 * 10$$

$$E = 516 \text{ VDC and the short circuit current of 5.61 amps}$$

The size of the circuit conductors must be done in accordance with Articles 690.8(A-1) and (B-1). In accordance with A-1 the sum of the short circuit current of the parallel-connected modules must be multiplied by 125%. In Figure 2-13 the short circuit current is 5.61 A, therefore, the short current (Isc) is multiplied by 1.25(Isc * 1.25), so the following is 5.61 A * 1.25 = 7.01 A. But the code goes further in B-1, stating that the overcurrent devices and the conductors are sized to carry not less than 125% of the maximum currents as calculated in Article 690.8(A). Therefore the 7.01 calculated in A-1 is multiplied by 125% (7.01 A * 1.25 = 8.76 A).

## Conductors Size

To determine the correct wiring size for the photovoltaic source circuit, we must start with the short circuit current and open-circuit voltage. In Figure 3-14 the corrected current is 8.76 A and the voltage at the photovoltaic source DC voltage is 516 volts DC. Then the ambient temperature, conduct fill, and the voltage drop must be factored.

A factor that must be dealt with is ambient temperature when the conductors are run with the array temperatures 70°C or higher. This means that the wiring, and any nonmetallic raceway if used, must have a minimum of a 90°C rating. For example, the 8.76 A load, in Figure 3-14, the #14 AWG THWN-2 has a current capacity of 25 A on Table 310.16 for 70°C. After applying a de-rating factor of 0.58 (from Article 310.16) this would be 14.5 A, which will work for the 8.76 A load.

Voltage drop should always be reviewed at this point. For example, if the array is 200 feet from the PV to the combiner box and is carrying 9 amps at 516 VDC and the wire size is #14 AWG. $V_d = 2 \times K \times L \times I/CMIL$, we know from Chapter 1 that $K = 12$ for copper wire, L is the one-way length of the cable run 200 feet, and I = adjusted short circuit current is rounded up to 9 A. The voltage drop $V_d$ is calculated as:

$$V_d = \frac{2 \times 12 \times 200 \times 9}{4110}$$

$$V_d = \frac{43,200}{4110}$$

$V_d = 10.51$ volts

$$V_d\% = \frac{V_d}{V}$$

$$V_d\% = \frac{10.51}{516}$$

$$V_d\% = 2.04\%$$

This would be too high for this segment of the circuit; it would be better to increase the size to #10AWG—the savings in kWs will more than pay for the increase in wire costs.

The voltage drop $V_d$ is calculated as:

$$V_d = \frac{2 \times 12 \times 200 \times 9}{10,380}$$

$$V_d = \frac{43,200}{10,380}$$

$V_d = 4.16$ volts

$$V_d\% = \frac{V_d}{V}$$

$$V_d\% = \frac{4.16}{516}$$

$$V_d\% = 0.8\%$$

The voltage drop of 0.8% is acceptable for this segment of the system. We will review this when we calculate the remainder of the system. The NEC states that 3% is the total voltage drop for the entire system; it is important to keep this number as low as possible—it will cost more in energy losses over the life of the system.

In Figure 3-14 there are 24 strings connected in parallel at the Central String-Monitoring System for an open-circuit voltage of 516 VDC and a short circuit current of 33.66 amps. Corrected for Articles 690.8 (A-1) and (B-1), the correct circuit current to size the conductors from the combiner boxes would be 52.59 amps—the correct wire size from Table 6 AWG. If we say that the inverter is at a distance of 150 feet, the voltage drop would be 1.3%.

### Wiring Methods

The wiring methods must meet the requirements of Article 690.31(A). This article of the code allows for the use of all raceways and cable wiring methods in the code and those other wiring systems and fittings that are listed and labeled for use in photovoltaic arrays shall be permitted. Article 690.31(A) states that where photovoltaic source and output circuits that have maximum operating system voltages greater than 30 volts are installed in readily accessible locations, circuit conductors shall be installed in a raceway.

Article 690.43 requires that the equipment-grounding conductors for the photovoltaic array shall run with the photovoltaic circuit either contained in the raceway or cable, or run with the conductors when they leave the vicinity of the array.

## 3 – String Monitoring System

Sunny Central String-Monitoring System (SMS) replaced the standard combiner box (Figures 3-15 and 3-16). The SMS allows for the convenient termination of the

Courtesy of SMA Solar Technology AG

■ **Figure 3-15** Sunny Central string-monitoring system. (See detailed specifications at http://www.sma-america.com.)

| | Sunny Central String-Monitor US Version 24 | Sunny Central String-Monitor US Version 32 | Sunny Central String-Monitor US Version 64 |
|---|---|---|---|
| **PV generator connection** | | | |
| Input voltage range | 0 ... 600 V DC | 0 ... 600 V DC | 0 ... 600 V DC |
| Max fuse size (10 x 38 class CC fuses) | 20 A, 600 V DC [1] | 15 A, 600 V DC [1] | 8 A, 600 V DC [1] |
| Max. PV short-circuit current per string | 12.8 A [2] | 9.6 A [2] | 5.1 A [2] |
| Max. number of strings | 24 | 32 | 64 |
| Fused inputs per measuring channel | 3 | 4 | 8 |
| PV array configuration | neg. or pos. grounded | neg. or pos. grounded | neg. or pos. grounded |
| Number of measuring channels | 8 | 8 | 8 |
| **Sunny Central Connection** | | | |
| DC short-circuit current | 480 A | 480 A | 512 A |
| Max. operating output current, A DC, continuous | 308 A | 308 A | 328 A |
| Max. number of cables per output port | 2 | 2 | 2 |
| **Mechanical Data** | | | |
| Dimensions: W x H x D in inches | 31.5 x 31.5 x 9.8 | 31.5 x 31.5 x 9.8 | 31.5 x 47.2 x 11.8 |
| Weight | 143 lbs | 146 lbs | 194 lbs |
| Protection rating | NEMA 3R | NEMA 3R | NEMA 3R |
| Housing material | Steel or Aluminum | Steel or Aluminum | Steel or Aluminum |
| **Ambient conditions** | | | |
| Permissible ambient temperatures | –13 °F to 113 °F | –13 °F to 113 °F | –13 °F to 113 °F |
| Rel. humidity | up to 95%, condensation possible | up to 95%, condensation possible | up to 95%, condensation possible |
| **Communication** | | | |
| Connection SSM-US | RS485 | RS485 | RS485 |

1) These values indicate input current without the derating factor of NEC articles 690.8(A)(1) and 690.8 (8)(1) applied.
2) These values indicate input current with the derating factor of NEC articles 690.8(A)(1) and 690.8 (8)(1) applied.

| Type designation | SSM-US | SSM-US | SSM-US |
|---|---|---|---|

## SSM-US Version 32, Negative Grounded

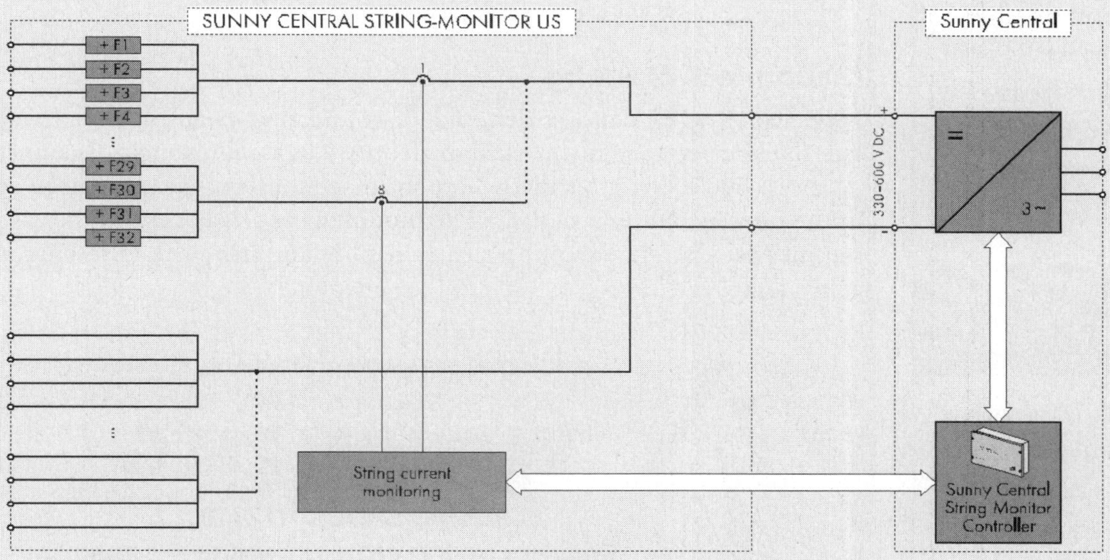

**Tel. +1 916 625 0870**
**Toll Free +1 888 4 SMA USA**
**www.SMA-America.com**

**SMA America, LLC**

Courtesy of SMA Solar Technology AG

**Figure 3-16** Sunny Central monitoring system specifications. (See detailed specifications at http://www.sma-america.com.)

photovoltaic source circuits. It contains the fuses for the photovoltaic circuit and provides the disconnect means for the photovoltaic strings. It also provides a way to monitor the output of each string in the PV system. The processor of the system monitors each string and sends the data via RS-485 link to a central point where the data can be monitored using SMA's system or the client's system. The data will be used to identify faulty strings, then the maintenance personnel can diagnose the problem with the string. This could be as simple as dirt on the panels or a defective panel. This is very important in any system, especially a large PV system. See Figure 3-14.

### Disconnecting Requirements

NEC 690.13–18 sets forth the requirements for the disconnect means that are required for photovoltaic systems.

NEC 690.13 requires that there shall be a means to disconnect all current-carrying conductors of the photovoltaic circuit from all other conductors in a building or other structure. It also states that a switch, circuit breaker, or other disconnect device—either AC or DC—be installed in the ungrounded conductor if the operation of the device leaves the marked, grounded conductor in an ungrounded and energized state. The grounded conductor may have a terminal or bolted connector that allows qualified personnel to perform maintenance or troubleshooting of the equipment. This is very important as this can cause physical injury and damage to the photovoltaic equipment.

There is an exception to the requirements of Article 690.13. If the switch or circuit breaker is part of the ground fault detection system required by Article 690.5, it shall be permitted to open the grounded conductor of the photovoltaic circuit when the device is automatically opened as a normal function of the device in responding to ground faults. The device shall indicate the presence of a ground fault.

### Disconnect Marking

NEC 690.14 (C-2) requires that each disconnecting means in a photovoltaic system shall be permanently marked to identify it as a photovoltaic disconnect.

NEC 690.17 requires that where the disconnecting means may be energized in the open position, a sign shall be mounted on or adjacent to the disconnecting means. The sign shall be clearly legible and have the following words or equivalent:

**Safety First**

Care must be taken when working with photovoltaic source circuits. Never disconnect the ungrounded circuit under load and never remove the fuse under load. This will cause DC arcing, which can cause physical injury and damage the fuse holder and circuit buss. It is best to cover the array and disconnect the circuit from the load whenever working on the system. Care should also be taken when working with a processor in the system; short circuiting the input can damage the system.

**Safety First**

It is important for the safety of others that you clearly mark the disconnect means. As the installer, you are aware of the safety hazard. When the system is serviced sometime in the future, it is important to warn the servicing personnel.

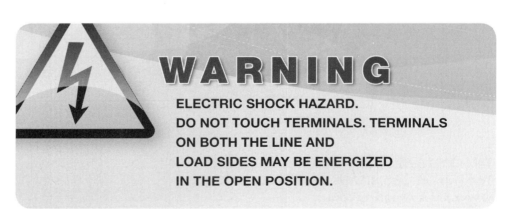

**WARNING**

ELECTRIC SHOCK HAZARD.
DO NOT TOUCH TERMINALS. TERMINALS
ON BOTH THE LINE AND
LOAD SIDES MAY BE ENERGIZED
IN THE OPEN POSITION.

### Maximum Number of Disconnects

NEC 690.14(C-4) states that the maximum number of disconnecting means in a photovoltaic system should not exceed six switches or circuit breakers. The system in Figure 3-14 uses the SMS with 12 fused switches; therefore, we must provide a disconnect means for each array of strings in the system.

### Fuses

Fuses in the SMS are energized from both sides when the circuit is energized. This is due to that fact that the photovoltaic source circuits are connected in parallel and the fuses are fed from the buss and the circuit connected. Article 690.16 requires that a disconnecting means be provided to disconnect a fuse from all sources of the supply if the fuses are accessible to other than qualified persons. Such a fuse in the photovoltaic source circuit shall be disconnected independently of fuses in other circuits. In the SMA SMS shown in Figure 3-14 there are finger-safe fuses that disconnect the circuit without allowing the personnel to touch the fuse / fuse holder. Also there is a separate fuse for each photovoltaic string.

### Fuse Sizing

In an electrical system, fuses are used to protect the wiring and equipment from excessive currents. If these currents are allowed to flow, they will cause damage, heating, or in the extreme, fire. If the fuse is too small, it could open during normal operation. If the fuse is too large, it cannot provide the required protection.

The minimum sizes of the fuses are calculated using the short circuit current (Isc) of the photovoltaic source circuit. The NEC 690.8(A and B) requires that all fuses be sized for a minimum of 1.56 times the Isc of the circuit. In the photovoltaic source circuit in Figure 3-14 the Isc = 5.61 Adc, then we would determine the fuse size by Fs = 5.61 * 1.56, Fs = 8.75. The next standard fuse size would be 10A 600 VDC fuse.

## 4 – Data Link Circuit

The SMS use an RS-485 3 wire data communication link to the central inverter (Figure 3-17). It is important to remember that the RS-485 circuit must be terminated on both ends. SMA provides a terminator to insert in the second socket of the last SMS in the system. See Chapter 1 for details on the RS-485 circuit.

## 5 – Photovoltaic Output Circuit

### Wire Size

The photovoltaic source circuits in the SMS are connected in parallel so the short circuit (Isc) is the sum of the Isc of the individual circuits. In the system in Figure 2-27 the Isc = Isc1 + Isc2 + Isc3 + Isc4 + Isc5 + Isc6, then Isc = 5.61 * 6 or Isc = 33.66 Adc. As required by 690.8(A and B) the Isc must be multiplied by 1.56, then Isc = 33.66 * 1.56 or Isc = 52.51 Adc. The wire size must be determined using Table 310.16. The wire size must be adjusted in the field for conditions, such as the mounting location of the inverter. If the conductors are exposed to sunlight, then they must be de-rated. Also conduit fill and voltage drop must be considered.

## 6.1  Preparing the Data Cable

Three wires of the data cable, GND (2), Data+ (3), and
Data– (4), are required for the connection. Data+ and
Data– must be a twisted pair.

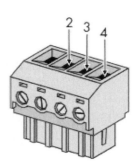

1.  Remove the insulation of the data cable by approx.
    2.4 in. (60 mm) and uncover the shield.

2.  Uncover the wires by approx. 1.8 in. (45 mm).

3.  Strip the wires by approx. 0.2 in. (1.8 mm).

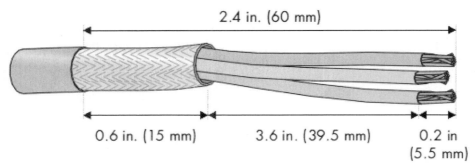

2.4 in. (60 mm)

0.6 in. (15 mm)    3.6 in. (39.5 mm)    0.2 in
(5.5 mm)

4.  Open the screw terminal completely.

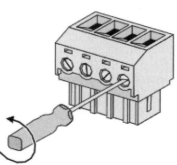

5.  Insert wires into the terminal.

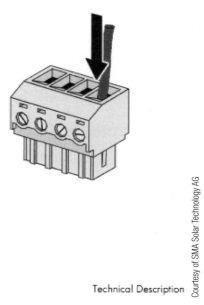

Courtesy of SMA Solar Technology AG

**Figure 3-17** RS485 data link connection. (See detailed specifications
at http://www.sma-america.com.)

82

## 6 – Inverter Inputs from other Arrays

The inverter used in Figure 3-14 has six inputs that can be used for additional arrays. Each array must have an RS485 circuit to the SMS in that system for monitoring of the array strings.

## 7 – Inverter

For complete details on the inverter, refer to Chapter 1. The inverter used in the example in Figure 3-14 has been designed with SMA inverters, which provides the required ground fault protection (GFP) and anti-islanding.

### Ground Jumper

NEC 690.49 requires that if the inverter in an interactive utility connected system is removed for maintenance, a bonding jumper shall be installed to maintain the system grounding while the inverter is removed. Bonding jumpers must be done in accordance with Article 250.120(C).

### Wire Size

NEC 690.8 requires that the maximum continuous output current of the inverter shall be used. The specification of the inverter must be consulted for the value. In the system in Figure 3-14 the maximum current (Im) is 300 A (see specifications in Appendix Figure A-7). As required by Article 690.8(A–B) the Im must be multiplied by 1.56, then Im = 300 * 1.56 or Im = 468 Aac. The wire size must be determined using Table 310.16. The wire size must be adjusted in the field for conditions, such as the mounting location of the inverter. If the conductors are exposed to sunlight, then they must be de-rated. Also conduit fill and voltage drop must be considered.

## 8 – Disconnect Means

NEC 690.14(C) states that means shall be provided to disconnect all conductors in a building or other structure from the photovoltaic system conductors.

In accordance with NEC 690.15, means shall be provided to disconnect the inverter(s) from all ungrounded conductors of all sources.

NEC 690.17 requires that where the disconnecting means may be energized in the open position, a sign shall be mounted on or adjacent to the disconnecting means. The sign shall be clearly legible and have the following words or equivalent:

**WARNING**

ELECTRIC SHOCK HAZARD.
DO NOT TOUCH TERMINALS. TERMINALS
ON BOTH THE LINE AND
LOAD SIDES MAY BE ENERGIZED
IN THE OPEN POSITION.

## 9 – Ethernet Network System to Monitor the Array Strings

The SMS can be monitored on an internal computer network or connected to a router and monitored over the network. It is important that the system is monitored on a regular basis to identify system faults.

## 10 – Ethernet

The inverter connects to the network via an Ethernet connection. See Chapter 1.

## 11 – Circuit breaker

The circuit breaker is the point of connection of the interactive inverter system. The rules of Article 690.45 must be followed for the connection to the main panel. In the system in Figure 3-14 the connection is made to the load side of the service entrance equipment and must follow Article 690.45(B1) through (B7).

1. It is required that each inverter or interconnecting system must have a dedicated fuse or circuit breaker.

2. The sum of the current rating of the overcurrent devices shall not exceed 120% of the rating of the buss bar or conductor.

3. The connection point shall be on the line side of the GPF equipment.

4. Equipment containing multiple devices with multiple sources shall be marked to indicate the presence of all sources.

5. If the circuit breakers are back fed, then the circuit breaker shall be suitable for such operation. If a circuit breaker is marked line and load, then it has been tested only in that direction.

6. Listed plug-in-type back fed circuit breakers from utility-interactive inverters complying with Article 690.60 shall be permitted not to have the additional fastener normally required by Article 480.36(D). This is allowed because if the breakers are lifted from the bus bar, then the inverter would trigger the anti-islanding function and shut off the AC output immediately as per Article 690.61.

7. The circuit breaker feeding the panel must be at the opposite end of the bus bar (load end) from the main breaker (line end) and be permanently marked with the following or equivalent.

**WARNING**
INVERTER OUTPUT CONNECTION
DO NOT RELOCATE THIS OVERCURRENT
DEVICE

# MICRO-INVERTER SYSTEM

A micro-inverter PV system is several AC modules within the total PV system—the separate AC modules that are strung together to form an array. Each module is an AC generating system that is a completely self-contained solar panel and inverter. This allows for an unlimited number of variations in system configurations. The designer can use different size and wattage panels that are needed to form the system. Also, each of the AC modules can be monitored separately allowing for maximum system efficiency and ease of maintenance.

NEC Article 690.2 defines an alternating current (AC) module as a complete self-contained unit with the inverter, solar panel, optics, and other components, exclusive of tracker, designed to generate AC power when exposed to light, see Article 690.2 for a complete list of definitions. NEC Article 690.6 sets forth the requirements for an AC module that must be adhered to. The system in Figure 3-19 is several AC-module grid-tied photovoltaic systems that uses one panel and a micro-grid-tied inverter, see Chapter 1. The inverters are mounted under the solar panels for ease of maintenance.

With an AC module system we do not calculate the system components, as we would do with a standard system. NEC Article 690.6(A–E) sets forth the requirements of the AC module. See Article 690.6.

(**F**) **Photovoltaic Source Circuits:** The requirements of Article 690 shall not apply to the connection of the inverter, source circuits, and components that are considered as internal wiring of the module.

(**G**) **Inverter Output Circuit:** The output of the AC module shall be considered an inverter output circuit.

(**H**) **Disconnect Means:** As required by Articles 690.15 and 690.17, a single disconnecting switch shall be permitted for the combined AC output of one or more AC modules. Each AC module in the system shall be provided with a connector, bolted, terminal-type disconnecting means.

**Safety First**

As the AC module will be generating an output as soon as the modules are installed, care must be taken to prevent accidental exposure to electrical current.

Courtesy of Enphase Energy, Inc.

■ **Figure 3-18** Enphase energy M190 micro-inverter. (See detailed specifications at http://www.enphaseenergy.com.)

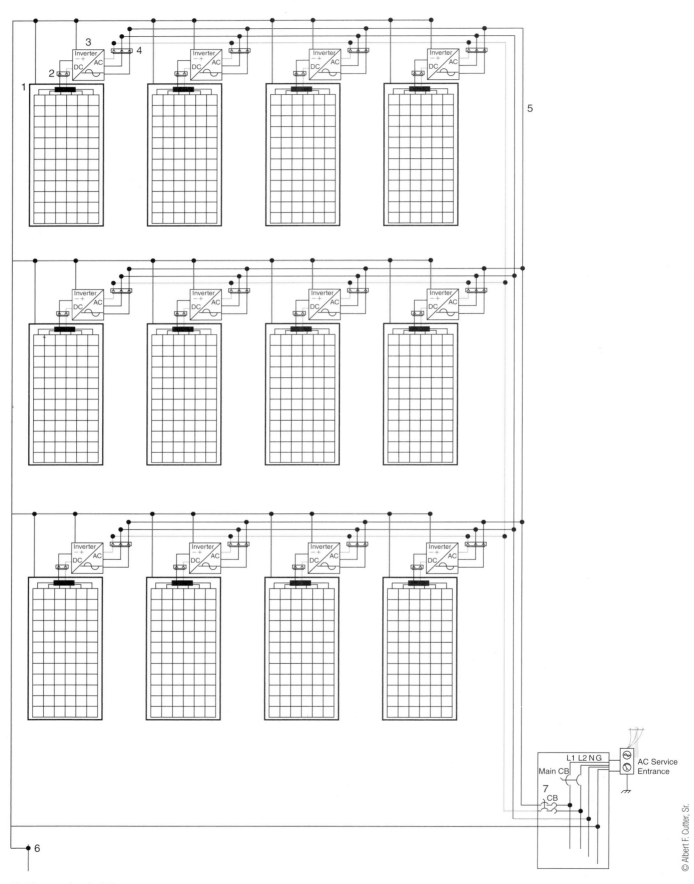

■ **Figure 3-19** Micro-inverter array.

**(I)   Ground-Fault Detection:** AC module systems shall be permitted to have a single detection device to detect AC ground faults and disconnect the AC power from the module.

**(J)   Overcurrent Protection:** The output circuits of an AC module shall be permitted to have conductor sizing and overcurrent protection in accordance with Article 240.5(B2).

NEC Article 690.52 requires that the AC module be marked with identification of terminals or leads and with the following ratings:

A—Nominal operating AC voltage

B—Nominal operating AC frequency

C—Maximum AC power

D—Maximum AC current

E—Maximum AC module overcurrent protection device rating

The following is a description of Figure 3-19 using the numbered points in the circuit.

1—In the example in Figure 3-19 we have a photovoltaic panel, for the purposes of this discussion we will use the Sanyo 215-watt panel with the specifications in Figures 2-12 and 2-13. This will mean that we have a 215-watt panel with open-circuit voltage of 51.6 volts output at short circuit current of 5.61 amps DC. The maximum circuit voltage should still be calculated for the lowest ambient temperature.

Voltage temperature coefficient (Vtc) for this panel would be $-V/^{\circ}C$; using an ambient temperature of $-4^{\circ}C$, we can calculate the Vmax = 53.447 VDC.

This voltage shall be used to calculate the voltage rating of the cables, disconnects, overcurrent devices, and other equipment that may be used.

The maximum current is the short circuit current of all the modules in parallel as setforth in Articles 690.8(A)(1) and (B)(1); the multiplication factor will be 156%.

Imax = Isc * 156%

Imax = 5.61 * 1.56

Imax = 8.75 amps

2—The AC module is not required to have connectors between the solar panel and the inverter but it is usually provided for servicing of the system. If a connector is used, it cannot disconnect the equipment-grounding conductor and it must meet the requirements of NEC Article 690.33. This connector will have the same connector that is used on the PV panel. See Appendix Figures A-1 and A-2.

3—An inverter is used to convert the DC output of the photovoltaic module to AC voltage required by the grid, see Chapter 1. NEC Article 690.2 defines this system as an interactive system, which requires that it have anti-islanding capability and ground-fault protection, which is usually built into the inverter. Always check the inverter specifications.

4—In accordance with Article 690.6(C) the system shall be provided with a connector, bolted, or terminal-type disconnected means.

5—The system must provide a means of disconnecting from the AC power of the utility grid and the requirements of NEC Articles 690.6(C), 690.15, and 690.17 must be met. Also Article 690.54 requires that the following markings be made at an accessible location at the disconnection means:

A—Rated output AC current

B—Nominal operating AC voltage

6—Grounding

This system is an AC module with the grounding taken care of in the assembly of the module. The system has a DC open-circuit voltage of 51.6 volts and therefore requires that the DC system be grounded. See NEC Article 690.41, which requires that and a DC source system over 50 volts must be grounded, which means that one conductor must be grounded. A smaller panel can be used in the 100-watt range, which would have been under 50 volts. The panel in Figure 2-11 is used as we have the specifications. Also due to the high efficiency of the SANYO HIT panel, it makes sense to use the highest output panel. NEC Article 690.42 requires that this shall be made at a single point on the photovoltaic output circuit. This is required so that the ground is not supplied from different sources if the ground is broken or removed during maintenance procedures. The inverter has built-in GFP and the grounding of the DC and AC circuit will happen at the GFP internal to the inverter.

Equipment-Grounding Article 690.43 requires that exposed noncurrent-carrying metals parts of the photovoltaic panels, metal frames, equipment, and conductor enclosures shall be grounded in accordance with Article 250.134 or 250.136(A). The article continues to talk about grounding photovoltaic panel frames to each other. Since we are working with a single AC module in this example, it does not apply.

The size of Equipment-Grounding Conductors are covered in Article 690.45 for the system in Figure 3-19. The article states that the conductor be sized using the open-circuit current, which is 5.61. For this example, using Table 250.122, the smallest size is #14 AWG. The article further requires that the grounding conductor shall not be smaller than #14 AWG.

Array Equipment-Grounding Conductor Article 690.46 states that equipment-grounding conductors smaller than the #6 AWG shall comply with Article 250.120(C), which states that the a grounding conductor smaller than the #6 AWG shall be protected by a raceway or armored cable or where not subject to physical damage. In the AC module of this example we can say that the ground conductor is protected from damage, as it is part of the integral wiring system of a single-panel AC module.

The Grounding Electrode System Article 690.47 covers the system in Figure 3-19, which is an AC module system so Article 690.47(A) applies to Figure 3-19. The grounding electrode system shall comply with Articles 250.50 through 250.60. Grounding Electrode System Article 250.50 states that all electrodes described in Articles 252(A)(1) through (A)(7) shall be bonded together.

Article 690.47(D), Additional Electrodes for Array Grounding, requires that the grounding electrode shall be installed at the location of ground- and pole-mounted systems and as close as possible to roof-mounted systems.

Summary of equipment grounding:

Grounding conductor electrode shall be no smaller than #14 AWG and must be protected from damage by a raceway or armored cable. The grounding conductor must be continuous—the conductor cannot be spliced or broken. The grounding electrode shall be installed in accordance with Article 250.52 at the location of the system.

7—Inverter Output Connection

The panel must be rated greater than the sum of ampere ratings of all the over-current devices supplying the panel. A connection to the panel shall be positioned at the opposite load end from the main circuit or feeder location. See Article 690.64.

A permanent warning label shall be applied to the distribution equipment with the following or equivalent marking:

**WARNING**

**INVERTER OUTPUT CONNECTION**
**DO NOT RELOCATE THIS OVERCURRENT**
**DEVICE**

The system is not required to have system monitoring. But it is essential for the requirements for ease of maintenance and maximum efficiency. The system in Figure 2-32 uses a PLC system. This will monitor critical sensor points like photovoltaic DC output voltage, current, ambient temperature, and inverter AC output voltage. This data is transmitted over the power line to a remote monitoring center where the information can be used to monitor the performance of the unit, see Chapter 1.

## REVIEW

1. Ohm's Law is basically stated as E=I/R.
   A. True
   B. False

2. A PV cell's voltage is at its highest when no current is being drawn.
   A. True
   B. Flase

3. A PV cell's current is at its maximum when no load is present.
   A. True
   B. False

4. If a PV cell has a voltage of 0.5828 and the wattage is 3.25 watts, then what is the current?

5. The PV modules maximum power (Ipm) = 5.13 amps and a short current of 5.61 amps. What value shall be used to calculate the equipment requirements?

6. Photovoltaic cells in the module are connected in series. There are 72 cells with a voltage of 0.5526 per cell. What is the total voltage of the module?

7. What is your number one responsibility when working with a PV system?

8. What does Article 690.4 mandate?

9. Which NEC 2011 article sets forth the requirements of an AC module?

10. The maximum current of an AC module system is the short circuit current of all the modules connected in parallel as set forth in Articles 690.8(A)(1) and (B)(1). If a short circuit current is 5.61 amps. What is the current used to calculate the equipment (Imax)?

11. In Figure 2-17 what does PLC stand for?

12. Define a monopole subarray.

13. A string array has multiple monopole arrays connected to a single inverter.
    A. True
    B. False

14. With a current of 8 amps and a voltage of 420 volts DC. Assume a #12 copper wire at a length of 550 feet is being used. What would be the voltage drop expressed in volts?

15. A central inverter system has multiple PV subarrays connected to a combiner box and a single inverter.
    A. True
    B. False

# CHAPTER 4

# WIND POWER

## INTRODUCTION

The wind has been used a source of energy for over 3000 years. The early mariners used the wind to power the sails of their ships and in 200 BC windmills were used in China and Persia to power mills and water pumps.

Wind energy is actually a converted form of solar energy. The sun's rays heat the different parts of the earth at different rates. For instance, the land, sea, and forest absorb and reflect the heat at different rates. This in turn heats and cools portions of the atmosphere differently. The hot air rises, reducing the atmospheric pressure at the earth's surface, and the cooler air is drawn in to replace the warmer air and this causes wind.

Air has mass and, when in motion, it contains energy. A portion of this kinetic energy can be converted into mechanical energy. This mechanical energy can be used for many practical applications like pumping water, sawing wood, and producing electrical energy.

In the winter of 1887 Charles F. Brush constructed the first automated wind power generator in North America. As you can see in Figure 4-1, it was a massive structure (note the man cutting the lawn). The 17-meter, 144-bladed rotor produced 12 kilowatts. It was the first windmill to incorporate a step up gearbox, which had a gear ratio of 50:1. This turned a direct current generator at the required speed of 500 rpm.

Modern commercial wind turbines have rotors from 8 to 110 meters in size and can produce from 250 watts to 5 megawatts (MW). The output of the wind turbine depends on the size of the rotor and the speed of the wind passing through it. The electrical systems that are used by wind turbines are basically the same for most applications.

In Figure 4-2 the wind rotor turns the alternator at varying speeds. This produces a variable-frequency, variable-voltage output from the alternator, as shown in Figure 4-3. This is commonly called "Wild AC." The Wild AC is full of harmonics and cannot be used directly and must be conditioned. To do this, it is fed

In this chapter the reader will obtain an understanding of the electrical systems used in wind turbine systems today.

Source: Wikipedia

■ **Figure 4-1** Charles F. Brush's wind generator 1887–1900.

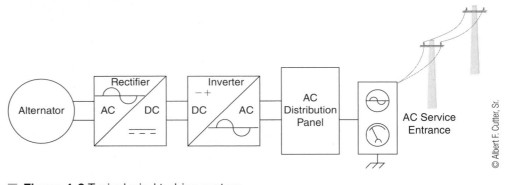

© Albert F. Cutter, Sr.

■ **Figure 4-2** Typical wind turbine system.

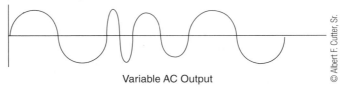

Variable AC Output

© Albert F. Cutter, Sr.

■ **Figure 4-3** Wild alternate current.

into a rectifier to convert it to direct current output as shown in Figure 4-4. Now the output of the rectifier produces conditioned DC power that can be fed into an inverter, which produces a constant-frequency, constant-voltage AC, as shown in Figure 4-5.

# LARGE WIND TURBINE

Figure 4-6 shows a 3-D view of the Nordex N100 2.5 MW wind turbine. You can clearly see the coupling, the gearbox, and the induction generator. On the backend of the housing you can see the weather station, which is used to determine the wind speed and wind direction. The many different types of wind turbines manufactured today will have a slightly different configuration but they will have the same basic components.

The Nordex N100 unit uses a double-fed asynchronous generator with a water-cooled cascade converter.

## Doubly-Fed Induction Generator System

The wind turbine in Figure 4-7 is a double-fed induction generator (WTDFIG) that is used in grid-tied turbine systems of all sizes. Each manufacturer has proprietary equipment to control and convert the output of the induction generator. In reality it is a rectifier and an inverter that converts the variable-voltage, variable-frequency AC to 50/60-cycle AC power at 475 to 690 volts AC. The wind rotates the turbine at a speed controlled by the wind speed and the pitch angle of the blades. The drive train gearbox converts the speed from the turbine to the rotation necessary to drive the generator. The stator and the rotor output the three-phase voltage to the converter, which is divided into two parts: C-rotor, the rectifier; and C-grid, the inverter. The microprocessor controller takes a signal from the C-rotor and C-grid to control the turbine pitch, thereby controlling the speed and the resulting power produced by the generator. A capacitor connected on the DC side of C-rotor acts as a DC source voltage. A coupling inductor L is used to connect C-grid to the output circuit.

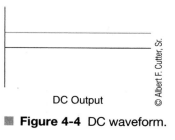

DC Output

■ **Figure 4-4** DC waveform.

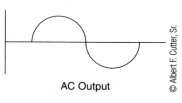

AC Output

■ **Figure 4-5** AC waveform.

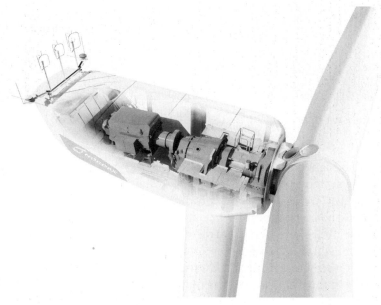

■ **Figure 4-6** Nordex 3-D view of a wind turbine.

© Courtesy of Nordex SE

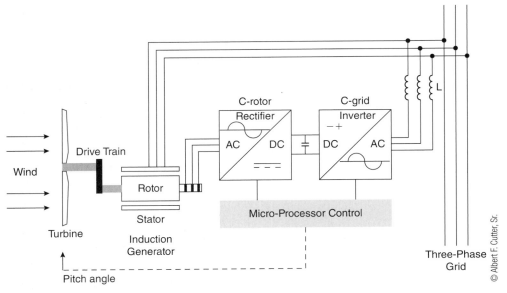

■ **Figure 4-7** Wind turbine doubly-fed induction generator system.

■ **Figure 4-8** Nordex wind farm in China.

Wind farms are designed for each location's unique physical characteristics and configuration of the system. Figure 4-8 shows a wind farm in China; Figure 4-9 shows a wind farm in Canada. They both produce electricity from the wind but with different physical characteristics and layout.

Figure 4-10, the Tehachapi Wind Farm, with about 5000 wind turbines, is the second largest collection of wind generators in the world (the largest is at the Altamont pass, near Livermore and the San Francisco Bay area), but it is now the largest wind power array in the world in terms of output. The turbines are operated by a dozen private companies, and collectively produce about 800 million kilowatt-hours of

■ **Figure 4-9** Nordex wind farm in Canada.

electricity—enough to meet the residential needs of 350,000 people every year. With over 15,000 turbines in the state (7000 at Altamont and 3000 at San Gorgonio Pass, near Palm Springs), wind power in California makes up about 1% of California's electricity.

■ **Figure 4-10** Tehachapi wind farm, California, United States.

## Large-Scale Wind Farm

Figure 4-11 is a typical wind farm which can have any number of turbines depending on the requirements and budget of the project. The proposed wind farm, which will be located 15–18 miles off the coast of New Jersey (United States), will have 96 turbines and produce 350 MW. Whether it is onshore or offshore, the system in Figure 4-11 is typical. Each wind farm will have its own unique configuration. The wind turbine induction generators will output a three-phase voltage in the range of 475 to 690 volts AC. This will be connected to a transformer alongside or in the base of the tower. This will transform the low voltage to medium voltage in the range of 25–40 kV; this range is typically used because it uses readily available competitively priced equipment. This is done to reduce the cost of the wire and reduce transmission losses. The medium voltage will be connected to Sub-Station A, where it will be stepped up to the voltage to connect to the grid. If the grid connection is a distance from the wind farm, then Sub-Station A will transform the medium voltage to high voltage (HVAC) in the range of 130–150 kV. Sub-Station B will transform the HVAC to whatever voltage is required by the grid. As with all complex systems, the system must be continuously monitored to maximize the efficiency of the systems. Large-scale wind turbines have monitoring and control systems that automatically control their efficiency and allow remote monitoring and control. The sub-station will have monitoring and control systems that will allow the system to be remotely monitored and continually adjusted as well.

## Small- to Medium-Scale Systems

Wind turbine systems up to and including 100 kW are covered under NEC Article 694, even if there are many 100 kW turbines connected in the system.

The majority of the small to medium turbines are direct drive permanent magnet generators. These turbines have less moving parts and complex systems and therefore require much less maintenance than the geared induction generator systems (Figures 4-12, 4-13, and 4-14).

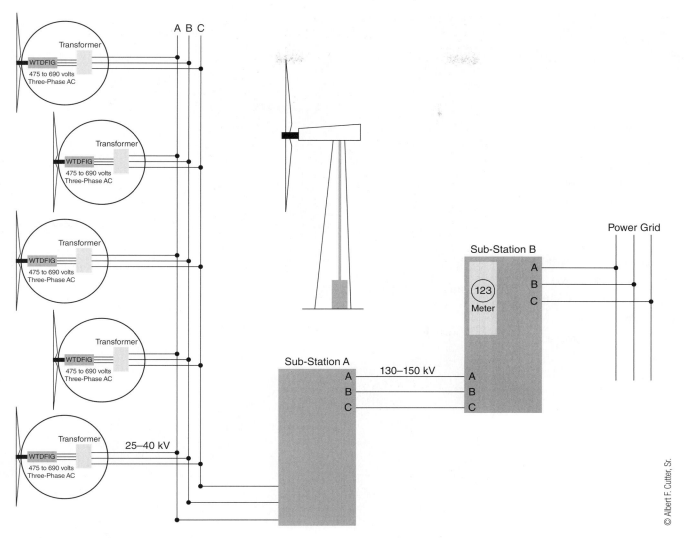

**Figure 4-11** Large-scale wind farm.

**Figure 4-12** Small wind turbine system.

© Courtesy of Barbara Cutter

■ **Figure 4-13** Small wind turbine.

© Courtesy of Fortis Wind Energy US

■ **Figure 4-14** Montana from Fortis Wind Energy. (See fortiswindenergy.us for complete specifications.)

## Single-Phase System

In Figure 4-15 we have a small 5 kW grid-tied wind turbine with an operating voltage of 80–400 volts and a maximum output of 468 volts, three-phase AC like the Fortis Wind Energy Montana 5 kW wind turbine shown in Figure 4-14.

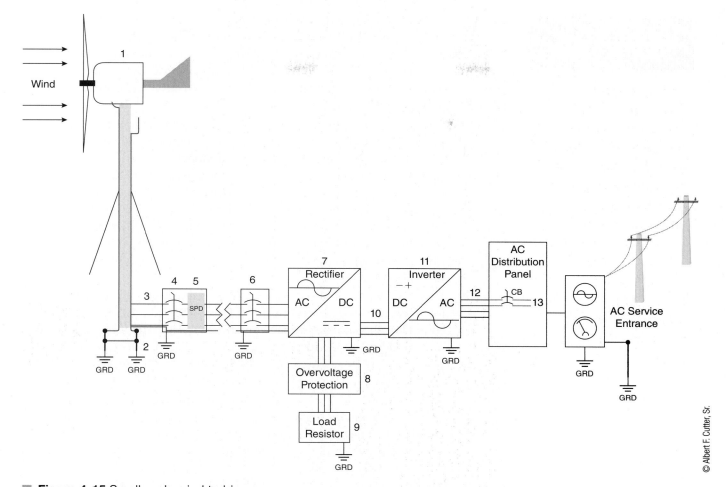

**Figure 4-15** Small scale wind turbine.

Article 694.10(A) requires that the maximum voltage for a wind turbine installed for a one- or two-family dwelling is 600 volts. Article 694.10(C) mandates that circuits over 150 volts must not be accessible to other than qualified personnel. This means that the system must be installed with lockable enclosures. According to Article 694.30(A) turbine output circuits over 30 volts that are installed in readily accessible locations systems must be installed in raceways.

Please refer to the numbered items on the single-phase system in Figure 4-15 for the following section:

1. The enclosure housing of the alternator and control circuits is called the nacelle as defined in Article 694.2 of the code. All exposed noncurrent-carrying metal parts of the turbine—including the nacelle, tower, and any other equipment—shall be grounded according to Article 250.134 or 250.136(A), as stated by Article 694.40(A). Care must be taken to remove any paint from the surface of the parts, and when connecting to aluminum surfaces use only rated lugs and an oxide-inhibiting joint compound to make the connection.

2. Tower grounding is covered under Article 690.40(C)(1–4) and should be read and understood before installing a tower. As stated in Article 690.4(C)(1),

**Safety First**

The noncurrent-carrying metal parts of the turbine must be grounded to protect against accidental electrical shock.

to limit voltages imposed by lightning, auxiliary electrodes shall be used. It is important that if grounding electrodes are used in close proximity to a galvanized foundation or galvanized tower anchoring components, then a galvanized grounding electrode shall be used. This is very important because copper or copper-clad electrodes can cause electrolytic corrosion of the galvanized components.

3. Circuit Conductor Sizing—Article 694.12(A)(1) requires that the maximum current of the circuit shall be based on the circuit current of the wind turbine operating at maximum output power. Article 694.12(B)(2) requires that the circuit conductor and the overcurrent devices be sized to carry not less than 125% of the maximum current calculated in Article 694.12(A). In this case it would be 1.25 amps * 125% = 16 amps. Other factors must be taken into account when sizing the conductors, conduit fill, and ambient temperature. It is also important to calculate the voltage drop in the circuit to account for the length of the tower and the distance to the equipment.

4. Disconnect at the Tower—Article 694.20 requires a disconnect means for all current-carrying conductors of the wind turbine. The grounded conductor shall not be connected to the switch or circuit breaker; it must remain unbroken. A circuit disconnecting means is not required and should never be used for a wind turbine that uses the output circuit to regulate turbine speed. This could cause serious damage to the turbine.

5. Surge Protection Device (SPD) and Lightning Arrestor – A surge protector is a device that protects equipment from spikes in voltage that could damage the equipment. It is very important to protect the sensitive circuits of the components in the system. It is better that a SPD is blown versus a rectifier and possibly the inverter being destroyed. The SPD is mandated by Article 694.7(D).

6. Disconnect—The rectifier/inverter system is usually at a distance from the tower and in a building or enclosure. A disconnect switch is required at the converter system. This should be a disconnect switch with a circuit breaker rated at 125% of the circuit current, per Article 694.24.

7. Rectifier—The rectifier's function in the system is to take the variable-voltage, variable-frequency output of the turbine and convert it to DC for the inverter input. The device in Figure 4-16 is the SMA Windy Boy Protection Box (WBP) model WBP 600, which is used in Figure 4-15. This device has a maximum limit of 560 VDC. It is the function of the rectifier to protect the inverter from an overvoltage. A rectifier should not be connected without a proper load. The load resistor will be matched to the rectifier and wind turbine. The WBP box can be configured with an integral load resistor. Consult the manufacturer's specification for proper size and type.

8. Overvoltage Protection—The SMA WBP has overvoltage protection incorporated into its design. In other systems this can be a separate component and must be sized correctly for the system. Overvoltage protection is used to limit the voltage in case of a wind gust or a sudden storm in order to prevent damage to the inverter circuits.

9. Load Resistor—In the event the inverter voltage reaches a critical voltage, the turbine is loaded with the addition of a load or dump resistor. This is determined and controlled by the overvoltage protection device.

© Courtesy of SMA Solar Technology AG

■ **Figure 4-16** Windy Boy protection box. (See detailed specifications at http://www.sma-america.com.)

The load resistor is automatically connected when the voltage reaches a critical level. When the voltage drops below the critical level the resistor is disconnected; this limits the voltage to a safe level. Each inverter manufacturer will have resistors that work with their systems. Consult the manufacturer's specification for the correct type and size of resistor for the system.

10. DC Circuit—The size of the wire will be determined by current and the NEC. The input circuit conductors of the inverter shall be calculated at the maximum-rated input current of the inverter, per Article 705.60(A)(1). The recommended wire size for the SMA equipment is #10 AWG to #6 AWG 90°C (194°F) copper wires. Voltage drop may dictate that a larger size wire is required.

11. Inverter—The inverter used in the system in Figure 3-15 is a SMA Windy Boy. This inverter is designed for the unique operating characteristics of a small wind turbine. Article 694.60 requires that only inverters identified and listed as interactive shall be permitted in interactive systems. By definition, interactive inverters must have provisions for anti-islanding. If the service voltage fails or is outside of regulatory limits, the inverter must detect this and automatically disconnect from the service connection. This is done to protect the maintenance personnel that might be working down the line on the system service conductors.

**Safety First**

There is a risk of electrical shock working with the resistor in the system. All work on the system should be performed by qualified personnel only.

**Safety First**

Electrostatic discharges are possible when touching components inside the inverter. This may result to damage to the components. All electrostatic discharge provisions must be followed. Remove existing electrostatic charges by touching a grounded metal surface and wear static discharge equipment when possible.

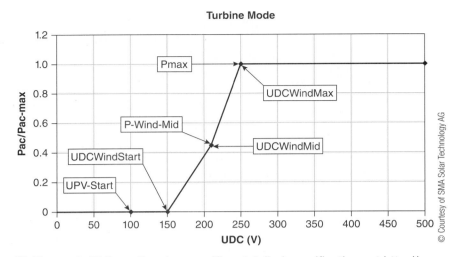

**Figure 4-17** Operational curve. (See detailed specifications at http://www.sma-america.com.)

The turbine mode of the Windy Boy inverter uses a programmable characteristic curve to regulate the input current depending on the generator voltage. Most small wind turbines use a permanent magnet generator. A characteristic of these wind turbine systems is that for every rotational/ wind speed there is a different optimal working point for voltage and current, this behavior is non-linear. Having a programmable curve allows the SMA Windy Boy inverter to match the operational curve of the generator for maximum efficiency and system performance.

The Windy Boy approaches this non-linearity using an approximation composed of a number of connected linear functions. These can be programmed by the user to correspond closely to the behavior of the small wind turbine system and thus provide performance adaptability. Figure 4-17 shows the connected linear functions of a typical characteristic curve for the Windy Boy.

12. Inverter Output Circuits – The maximum current of the circuit will be the continuous output current rating of the inverter, per Article 705.60(A)(2), and shall be sized to carry not less than 125% of the current rating of the inverter. An additional disconnect will be required if the panel circuit breaker is not in the line of sight of the inverter.

   An optional kW meter can be installed to monitor the turbine output. The inverter may also have a remote monitoring system that allows the user to monitor the turbine output.

13. The output of the inverter shall be considered continuous and the circuit breaker shall be sized at 125% of the rated output of the inverter, per Article 705.60(B). The breaker can be rounded up to the next standard size as provided by Articles 240.4(B) and (C).

## Three Phase System

The system in Figure 4-18 is a three-phase, 10–25 kW, grid-tied wind turbine system with an operating voltage of 80–400 volts and a maximum output of 468 volts three-phase AC. According to Article 694.30(A) turbine output circuits

**DANGER**

Never connect a load between the inverter and the circuit breaker of the loadcenter. This will cause an overload that cannot be detected by the loadcenter circuit breaker.

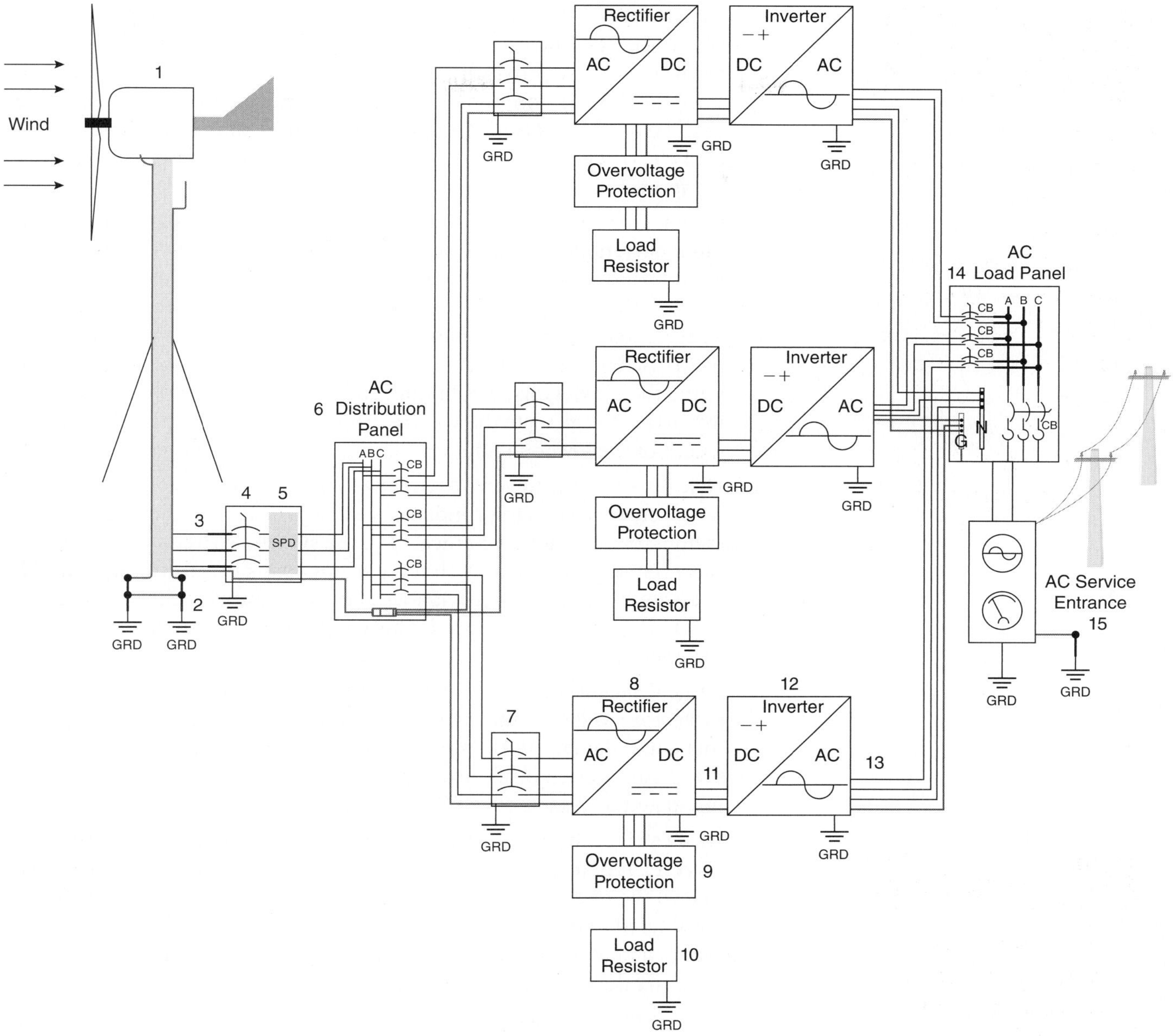

**Figure 4-18**

over 30 volts installed in readily accessible locations systems must be installed in raceways.

Please refer to the numbered items on the three-phase system in Figure 4-18 for the following section:

1. All exposed noncurrent-carrying metal parts of the turbine including the nacelle, tower, and any other equipment shall be grounded according to Article 250.134 or 250.136(A), as stated by Article 694.40(A). Care must be taken to remove any paint from the surface of the parts and when connecting to aluminum surfaces use only rated lugs and an oxide-inhibiting joint compound to make the connection.

2. Tower grounding is covered under Article 690.40(C) 1–4 and should be read and understood before installing a tower. As stated in

Article 690.4(C)(1), to limit voltages imposed by lightning, auxiliary electrodes shall be used. It is important that if grounding electrodes are used in close proximity to a galvanized foundation or galvanized tower anchoring components, then a galvanized grounding electrode shall be used. This is very important because copper or copper-clad electrodes can cause electrolytic corrosion of the galvanized components.

3. Circuit Conductor Sizing—Article 694.12(A)(1) requires that the maximum current of the circuit shall be based on the circuit current of the wind turbine operating at maximum output power. Article 694.12(B)(2) requires that sizing of the circuit conductor and the overcurrent devices be sized to carry not less than 125% of the maximum current calculated in Article 694.12(A). All other factors must be taken into account when sizing the conductors, conduit fill, and ambient temperature. It is also important to calculate the voltage drop in the circuit to account for the length of the tower circuits and the distance to the equipment.

4. Disconnect at the tower—Article 694.20 requires a disconnect means for all current-carrying conductors of the wind turbine. The grounded conductor shall not be connected to the switch or circuit breaker; it must remain unbroken. A circuit disconnecting means is not required and should never be used for a wind turbine that uses the output circuit to regulate turbine speed. This could cause serious damage to the turbine.

5. Surge Protection Device (SPD) and Lightning Arrestor—A surge protector is a device that protects equipment from spikes in voltage that could damage the equipment. It is very important to protect the sensitive circuits of the components in the system. The SPD is mandated by Article 694.7(D).

6. AC Distribution Panel—The distribution panel is required to take the three-phase output of the turbine and split it into three smaller circuits to feed the rectifiers/inverters. The input circuits of the rectifiers are considered to be continuous and should be rated at 125% of the rate input current of the inverter. Voltage drop must be considered when selecting the conductor size.

7. The rectifier/inverter system is usually at a distance from the tower and in a building or enclosure. Therefore a disconnect switch is required at the converter system. This should be a disconnect switch with a circuit breaker rated at 125% of the circuit current (Article 694.24).

8. Rectifier—The rectifier's function in the system is to take the variable-voltage, variable-frequency output of the turbine and convert it to DC for the inverter input. (See Figure 3-16.)

   The device in Figure 4-16 is the SMA Windy Boy Protection Box (WBP) model WBP 600, which is used in Figure 4-15. This device has a maximum limit of 560 VDC. Its function is to protect the inverter from an overvoltage condition. A rectifier should not be connected without a proper load. The load resistor will be matched to the rectifier and wind turbine. The WBP box can be configured with an integral load resistor. Consult the manufacturer's specification for proper size and type of load resistor.

9. Overvoltage Protection—The SMA WBP has overvoltage protection incorporated into its design. In other systems this can be a separate

component and must be sized correctly for the system. Overvoltage protection is used to limit the voltage in case of a wind gust or a sudden storm to prevent damage to the inverter circuits.

10. Load Resistor—In the event the inverter voltage reaches a critical voltage, the turbine is loaded with the addition of a load or dump resistor. This is determined and controlled by the overvoltage protection device. The load resistor is connected when the voltage reaches a critical level. When the voltage drops below the critical level the resistor is disconnected; this limits the voltage to a safe level. Each inverter manufacturer will have resistors that work with their systems. Consult the manufacturer's specification for the correct type and size of resistor for the system. The load resistor will generate heat; make sure that it is properly ventilated.

11. DC Circuit—The size of the wire will be determined by current and the NEC. The input circuit conductors of the inverter shall be calculated at the maximum-rated input current of the inverter, per Article 705.60(A)(1). The recommended wire size for the SMA equipment is #10 AWG to #6 AWG 90°C (194°F) copper wires. Voltage drop may dictate that a larger sized wire is required.

12. Inverter—The inverter used in the system in Figure 4-15 is a SMA Windy Boy. This inverter is designed for the unique operating characteristics of a small wind turbine. Article 694.60 requires that only inverters identified and listed as interactive shall be permitted in interactive systems. By definition interactive inverters must have provisions for anti-islanding. If the service voltage fails or is outside of regulatory limits, the inverter must detect this and automatically disconnect from the service connection. This is done to protect the maintenance personnel that might be working down the line on the system service conductors.

    The turbine mode of the Windy Boy inverter uses a programmable characteristic curve to regulate the input current depending on the generator voltage. Most small wind turbines use a permanent magnet generator. A characteristic of these wind turbine systems is that for every rotational/wind speed there is a different optimal working point for voltage and current, this behavior is non-linear. Having a programmable curve allows the SMA Windy Boy inverter to match the operational curve of the generator for maximum efficiency and system performance. (See Figure 4-17.)

13. Inverter Output Circuits—The maximum current of the circuit will be the continuous output current rating of the inverter, per Article 705.60(A)(2), and shall be sized to carry not less than 125% of the current rating of the inverter. An additional disconnect will be required if the panel circuit breaker is not in the line of sight of the inverter. An optional kW meter can be installed to monitor the turbine output.

14. The output of the inverter shall be considered continuous and the circuit breaker shall be sized at 125% of the rated output of the inverter, per Article 705.60(B). The circuit breaker can be rounded up to the next standard size as provided by Articles 240.4(B) and (C). The phase relationship of the inverters must be observed to ensure proper connection to the grid.

## REVIEW

1. Turbine systems up to 100 kW are covered by which article of the National Electrical Code?

2. Article 694.7 mandates that anyone can install a wind turbine system.
   A. True
   B. False

3. A surge protection device is used to protect the system from lightening.
   A. True
   B. False

4. Overvoltage protection is used to prevent the system from "Wild Current."
   A. True
   B. False

5. What is the purpose of the load resistor?

6. Anti-islanding is required to protect the maintenance personnel.
   A. True
   B. False

7. Electrostatic discharges from personnel working on the system cannot be prevented.
   A. True
   B. False

8. Three-phase systems use one large inverter.
   A. True
   B. False

# CHAPTER 5

# BATTERY SYSTEMS

## *objective*

This chapter will expose the reader to the battery systems for storage/backup of the alternative systems and UPS systems. By the end of this chapter the reader will have a basic understanding of battery systems and be able to work on these systems with an understanding of the system and relevant articles of the National Electrical Code (NEC).

## WHAT WE NEED TO KNOW

No one knows when the first battery was invented. Figure 5-1 shows a battery that was discovered in 1936 at an excavation at Khujut Rabu, near Bagdad, where several of these jars dating back to 250 BC were found. It is thought that the batteries were used for plating objects but then no one knows for sure what they were used for.

## INTRODUCTION

Benjamin Franklin, in 1748, strung together several Leyden jars and capacitors, and called the array a "battery." It was not until 1800 when Alessandro Volta, after extensive experimentation, invented the voltaic pile that we had the first battery of the modern age. The original voltaic pile consisted of zinc and silver discs, and between alternating discs, a piece of cardboard that had been soaked in salt water. A wire connecting the bottom zinc disc to the top silver disc would complete the circuit and produce a spark.

The problem with the voltaic pile was that it could not provide current for a sustained period of time. William Sturgeon, an electrical engineer, worked on the problem and in 1830 produced a battery with a longer life than that of Volta by amalgamating zinc.

One of the major problems with batteries was a process of polarization. This is a process in which a thin layer of hydrogen forms on the positive electrode, which increases the resistance of the battery, thereby reducing the effective electromotive force (emf).

John Daniell began experiments in 1835 in an attempt to improve the voltaic battery. His experiments led to remarkable results, and in 1836 he invented a primary cell in which hydrogen was eliminated in the generation of electricity. Daniell had solved the problem with polarization. He learned, in his laboratory, to alloy the zinc with mercury.

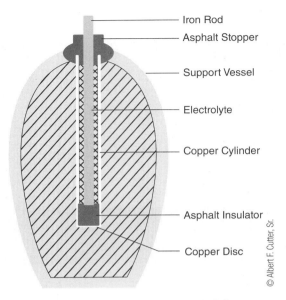

Iron Rod
Asphalt Stopper
Support Vessel
Electrolyte
Copper Cylinder
Asphalt Insulator
Copper Disc

© Albert F. Cutter, Sr.

**Figure 5-1** Baghdad battery 250 BC.

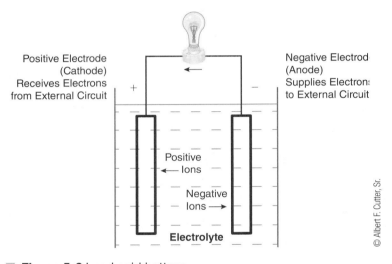

Positive Electrode
(Cathode)
Receives Electrons
from External Circuit

Negative Electrod
(Anode)
Supplies Electron:
to External Circuit

Positive
Ions

Negative
Ions

**Electrolyte**

© Albert F. Cutter, Sr.

**Figure 5-2** Lead acid battery.

In 1859 Gaston Planté began experiments that resulted in the construction of a battery for storage of electrical energy. A year later he presented a battery to the Academy of Sciences consisting of nine elements with sheets of lead, separated by rubber strips, rolled in a spiral, and immersed in a solution containing about 10% sulfuric acid. This battery could produce large amounts of current and became the first lead acid storage battery. Work continues today on improving on these batteries.

Shown in Figure 5-2 is a simple single-cell battery circuit. A battery is a device that produces electricity from a chemical reaction. It can be a single cell or many cells in series or parallel, but usually refers to a single cell. A cell consists of a negative electrode (anode); an electrolyte, which conducts ions; and a positive electrode (cathode).

There are two types of batteries: a primary cell, which is a battery that can only be used once then discarded; and a secondary cell, which is designed to be recharged and used many times. Secondary cells are used for storage batteries.

# LEAD ACID CHARGE/DISCHARGE CYCLES

A lead acid battery changes its chemical properties in the different states. In Figure 5-3 the battery is in the charged state. In this state the active materials consist of lead sponge on the negative plate and lead peroxide on the positive plate. The electrolyte is a mixture of sulfuric acid and water. The strength of the electrolyte is measured in terms of specific gravity of the mixture. This is the ratio of the weight of a given volume of electrolyte to an equal volume of water. Concentrated sulfuric acid has a specific gravity of 1.84; water has the specific gravity of 1.00. The specific gravity is different for each type and manufacturer of the cells. The readings used in this text are approximate figures, which are relative readings. In Figure 5-3 the hydrometer shows a reading of 1.280, this is the maximum value for this mixture, indicating that the cell is fully charged. In this state the active material of the positive plate is lead peroxide and the negative plate is lead sponge, and all the acid is in the electrolyte so the specific gravity is at its maximum value.

Figure 5-4 shows the cell in a state of discharge. In this state the acid, which is in the pores of the plates, separates from the electrolyte. This forms a chemical reaction

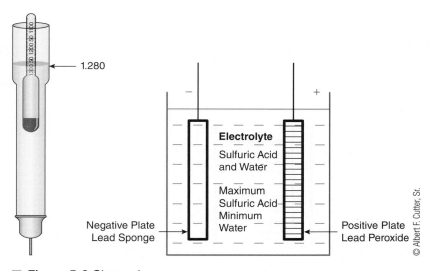

■ **Figure 5-3** Charged.

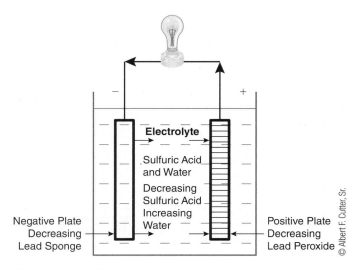

■ **Figure 5-4** Discharging.

with the active material lead changing to lead sulfate. As the cell continues to discharge, additional acid is drawn from the electrolyte into the pores and more sulfate is formed. As the cell is discharged, the specific gravity of the electrolyte will continue to decrease.

Figure 5-5 shows the cell completely discharged. In this state the positive plate has a minimum of lead peroxide and a maximum of lead sulfate, and the negative plate has a minimum of lead sponge and a maximum of lead sulfate. The hydrometer shows that the cell has a low specific gravity indicating that acid has been absorbed by the plates.

Figure 5-6 shows the cell in the charging state. As the cell charges the chemical reaction is reversed and the acid is drawn back into the electrolyte raising the specific gravity until it returns to the maximum level for this cell. The lead releases the acid, the positive plate returns to lead peroxide, and the negative plate returns to the lead sponge. This process will continue until the cell is fully charged.

Sealed lead acid batteries are used for backup power in many types of systems. Figure 5-7 is a sealed lead acid battery from APC. These batteries are low maintenance and are used in photovoltaic systems where there are many charge/discharge cycles. Figure 5-8 is a data center battery backup rack system from APC. Figure 5-8 is a complete APC battery backup rack system with integral charger.

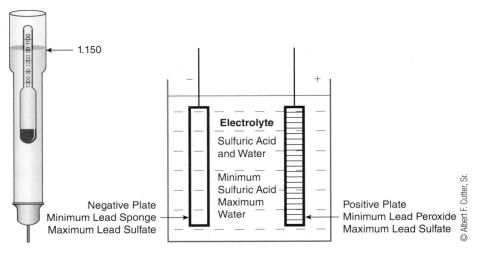

**Figure 5-5** Discharged.

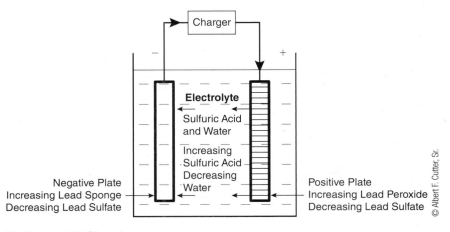

**Figure 5-6** Charging.

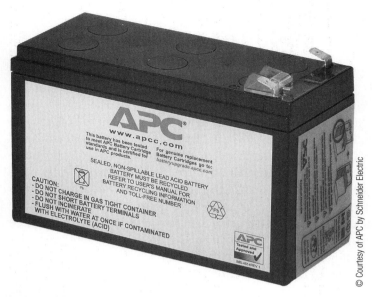

**Figure 5-7** Sealed Lead Acid Battery.

**Figure 5-8** Rack of sealed lead acid batteries.

**Figure 5-9** Rack of sealed lead acid batteries with charge controller.

# BATTERY WIRING DIAGRAMS

The following circuit diagrams illustrate the basic parallel, series, and parallel/series battery circuits. The parallel circuit is used to increase the circuit current, while the series circuit is used to increase the circuit voltage. The parallel/series circuit is used to increase the voltage and the current. It is important to note that the watts never change in the circuits. $P = E * R$, 2–12 volt batteries with 50 amps will have the power of 1200 watts no matter how they are connected. For the purposes of these circuits we use a 12-volt battery at 50 amps. A battery is typically rated in amp hours (Ah), which is the measure of the flow of current, in amperes, over one hour. A 350-Ah battery can produce 50 amps for 7 hours, 350/50 = 7 Ah.

Figure 5-10 shows a basic parallel battery circuit. In this configuration the current is increased but the voltage stays the same. The current is equal to $I_t = I_1 + I_2 + I_3$, assuming that the battery is rated at 50 amps the current would be $I_t = 50 + 50 + 50$, $I_t = 150$ amps. The voltage would equal $E_t = E_1, E_2, E_3$ or $E_t = 12$ volts. The watts or power of the circuit would be $P = E * I$, $P = 12 * 150$, $P = 1800$ watts. A circuit like this would be used to increase the backup time of a system without changing the circuit voltage.

Figure 5-11 shows a basic series battery circuit. In this configuration the current (I) stays the same throughout the circuit, $I = I_1, I_2, I_3$, and the voltage is additive, $E_t = E_1 + E_2 + E_3$. The current is $I_t = 50$ amps and the voltage equals $E_t = 12 + 12 + 12$, $E_t = 36$ volts. The power equals $P = E_t * I_t$, $P = 36 * 50$, $P = 1800$ watts.

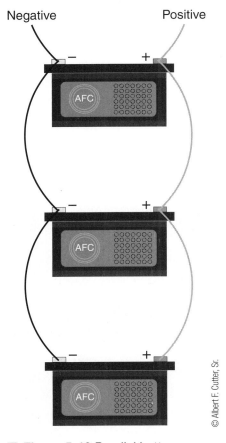

Negative                    Positive

© Albert F. Cutter, Sr.

■ **Figure 5-10** Parallel battery circuit.

Negative            Positive

■ **Figure 5-11** Series battery circuit.

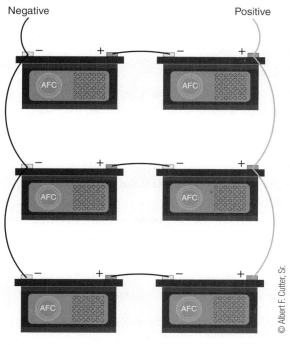

Negative       Positive

■ **Figure 5-12** Series/parallel circuit.

> **Safety First**
>
> Batteries emit hydrogen gas, which is very explosive; never use an open flame to check the electrolyte level.

> **Safety First**
>
> When charging, batteries emit hydrogen gas, which is very explosive. All tools should be insulated to prevent accidental short circuit, which will arch and could cause the battery to explode.

Figure 5-12 shows a basic series/parallel battery circuit. The voltage is equal to the voltage of column 1 + column 2. $E_t = Eb_1 + Eb_2$, $E_t = 12 + 12$, $E_t = 24$ volts. The current is additive of each row. $I_t = Ir_1 + Ir_2 + Ir_3$, $I = 50 + 50 + 50$, $I_t = 150$ amps at 24 volts. The power (watts) equal $P = E_t * I_t$, $P = 24 * 150$, $P = 3600$ watts.

# TYPES OF BATTERIES COMMONLY USED FOR BACKUP AND UPS SYSTEMS

- Lead Acid/Planté

  **Overview:** The battery is made of a lead grid framework into which lead paste or pure lead is applied. The plates are formed by charging with an electric current, which forms lead dioxide on the positive plate and porous (sponge) lead on the negative plate. Because the lead paste and grid framework is soft, special care must be exercised in the construction of these batteries.

**Cost:** 2 to 2.5 times the cost of lead acid/calcium batteries

**Maintenance:** Maintenance is low requiring little watering

**Number of deep discharges:** 100

**Life/Warranty:** 25 years

**Operating Considerations:** This battery can tolerate operation at high temperatures better then various lead alloy types, like anatomy and calcium. Because this battery generates hydrogen gas when charging and the electrolyte's sulfuric acid evaporates to some extent, these batteries must be used in a room that is well ventilated to the outside and kept away from delicate electronic equipment.

- Lead Acid/Antimony

  **Overview:** Lead antimony used in the grid construction strengthens the plates, which reduces construction cost.

  **Cost:** About the same as lead calcium

  **Maintenance:** Frequent water addition is required. Hydrogen gas generation and water usage is approximately ten times that of the lead acid/Planté and lead acid/calcium battery. There is also a need for periodic equalization.

  **Number of deep discharges:** 200

  **Life/Warranty:** 15 years; at the end of its life cycle it will lose approximately 20% of its original capacity.

  **Operating Considerations:** Because of the hydrogen, adequate ventilation is very important. This battery is selected when frequent discharging is expected.

- Lead Acid/Calcium

  **Overview:** Lead calcium alloy is used in the grid construction to strengthen the plates.

  **Cost:** Base price for comparison of other batteries

  **Maintenance:** Low watering is required

  **Number of deep discharges:** 200

  **Life/Warranty:** 20 years

  **Operating Considerations:** When maintained at a float voltage of 2.25 volts/cell it does not require routine equalization. As with most UPS systems, the battery will be on the charger floating most of the time and is not normally fully discharged. This battery is the most used in backup systems in the United States and offers the best cost versus service reliability. These batteries must be in a room with adequate ventilation.

- Lead Acid/Calcium, Maintenance-Free Liquid Electrolyte

  **Overview:** This battery is designed for 3–5 years of maintenance-free operation. It uses a high specific gravity electrolyte with a large electrolyte capacity.

  **Cost:** Approximately 35–50% the cost of lead acid/calcium batteries

  **Maintenance:** Little to none

  **Life/Warranty:** 1-year full replacement; 5-year life when used in a float application

**Operating Considerations:** Due to the gas generation, this battery requires adequate ventilation. Due to the fact that this battery is only manufactured in a limited number of sizes, it is necessary to parallel banks for larger back-ups or larger UPS systems. The price advantage is quickly lost when paralleling banks of smaller batteries.

- Lead Acid/Calcium, Maintenance-Free Gelled Electrolyte, Sealed

  **Overview:** This battery is a gelled, electrolyte-sealed lead acid/calcium unit that vents no gas under normal operation. The battery will vent gas as a safety precaution if the internal pressure reaches a specific level.

  **Cost:** 60–70% the cost of lead acid/calcium batteries

  **Maintenance:** No routine maintenance

  **Number of deep discharges:** N/A

  **Life/Warranty:** 1-Year full replacement; 20-year life

  **Operating Considerations:** The sealed battery can be safely used in an unventilated room making it suitable for applications where ventilation would be impractical or expensive.

- Lead Acid (Special Alloy), Suspended Electrolyte, Maintenance-Free, Sealed

  **Overview:** This battery is the newest of battery designs; it has an electrolyte suspended in a porous material. It is completely sealed and only vents gas as a safety precaution.

  **Cost:** 1 to 1.5 times lead acid/calcium batteries

  **Maintenance:** Maintenance-free

  **Number of deep discharges:** 400

  **Life/Warranty:** Up to 20 years.

  **Operating Considerations:** This battery can be used in an office environment or in an unventilated room. It has a high energy density and is physically smaller than an equal lead acid battery.

- Nickel Cadmium, Pocket Plate Liquid Electrolyte

  **Overview:** Nickel cadmium or NICAD batteries take 92 cells to equal 60 cells of lead acid cells; this battery still has the advantage of being smaller in size and weight for a given capacity. The NICAD battery does not experience the severe shorting of life when operated at elevated temperatures and performs better at low temperatures than lead acid batteries.

  **Cost:** 3 times the cost of lead acid/calcium batteries

  **Maintenance:** Low

  **Number of deep discharges:** 1000

  **Life/Warranty:** 20 years

  **Operating Considerations:** NICAD batteries do not emit hydrogen and oxygen gas, which are products of electrolysis, and there are no corrosive gases as with lead acid batteries. They can be installed directly next to delicate electronic equipment. When maintained at the recommended float voltage, periodic equalization is not required.

- Lithium ion

  **Overview:** Lithium ion batteries have four times the effective capacity at high discharge current versus lead acid batteries. More compact and lightweight units can be realized compared to lead batteries, which results

**Safety First**

Batteries can explode while charging when there are high amounts of hydrogen present. When working on batteries you must wear a face shield and protective clothing. It may be a little awkward to work in a safety suit but I have seen firsthand what can happen when sulfuric acid is splashed onto cotton clothing and bare skin.

in a small footprint. This battery has an energy density that is double or triple that of the lead batteries usually used as large-scale power sources; this results in a much smaller and lighter system, about half the volume and a third of the weight.

**Cost:** 2 times the cost of lead acid/calcium batteries

**Maintenance:** Low

**Number of deep discharges:** 2000

**Life/Warranty:** 20 years

**Operating Considerations:** Lithium ion batteries do not emit hydrogen and oxygen gas, which are products of electrolysis, and there are no corrosive gases as with lead acid batteries. They can be installed directly next to delicate electronic equipment. When maintained at the recommended float voltage, periodic equalization is not required.

# PHOTOVOLTAIC SYSTEM WITH BATTERY BACKUP

The system shown in Figure 5-13 is a photovoltaic system with battery backup. In the following sections we will look at points that must be considered.

## 1 - Photovoltaic Array

The solar arrays in Figure 5-13 are connected in parallel to form a 42-volt STC array. There are 10 panels in the array that are 200-watt each making this a 2 kW array. This is a 42-volt DC panel connected in parallel to produce a 42-volt DC array with an output current of 5.13—these are the STC values. To install and maintain the arrays we must use the open-circuit voltage of 51.6-volt DC and the short circuit current of 5.61-amp DC as required by the NEC 2011. This means that the unadjusted open-circuit voltage of the array is 51.6 volts DC. The panels are connected to the first combiner boxes with #12 AWG. This is sufficient to allow for voltage drop.

### *Installation and Maintenance*

NEC 690.18 states that short circuiting, open circuiting, and an opaque covering shall be used to disable the array or portions of the array.

### *Wiring Methods at the Array*

**Access to Boxes—**Boxes under the panels in 5-13 must be accessible; Article 690.34 is the rule that covers junction, pull, and outlet boxes. If the boxes are not directly accessible, boxes are considered accessible if the module or panel over them is secured with removable fasteners and the flexible wiring method is used. Ungrounded PV Power Systems

Photovoltaic systems are allowed to be ungrounded provided that they comply with Article 690.35(A)–(G). The system in 5-13 is a grounded system.

**Safety First**

Caution is needed when working with photovoltaic modules. When exposed to sunlight the modules will be energized and could expose the personnel to an electric shock or DC arcing.

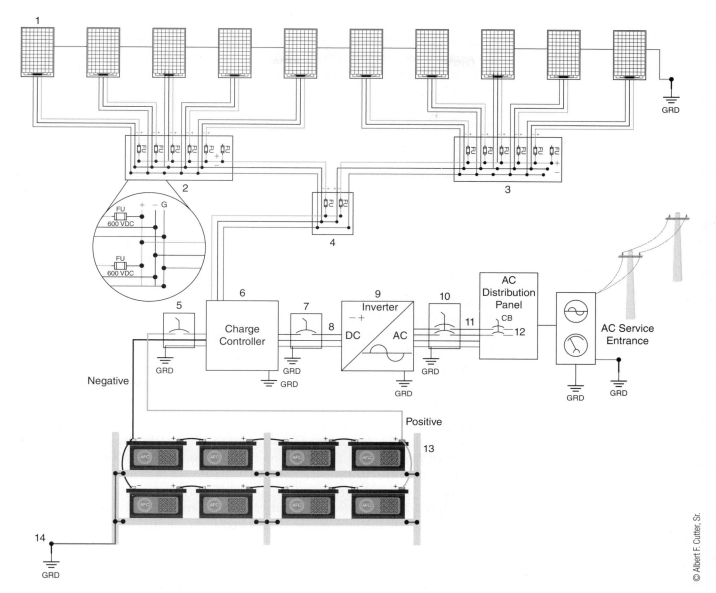

■ **Figure 5-13** PV system with battery backup.

**System Grounding**—The system in Figure 5-13 is over 50 volts. According to Article 69.41 the system must have a grounded circuit conductor. According to Article 690.42 the DC-grounding conductor must occur at only one point. There is an exception that requires that the connection correlates with the operation of the GFP (see Article 690.5), which in Figure 5-13 is at the inverter.

Now we must deal with the connection of the grounding electrode conductor, the equipment-grounding conductor. This connection is covered in Article 250.168 where an un-spliced bonding jumper must be installed at this point. This jumper must be sized the same way as the grounding electrode conductor for the system. According to Article 250.166(B) the grounding electrode conductor shall not be smaller than the largest DC circuit conductor supplied by the system, and not smaller than #8 AWG copper or #6 AWG aluminum.

**Equipment Grounding**—Article 690.43 requires that all noncurrent-carrying metal parts of the PV array conductors, enclosures, and support hardware shall be grounded in accordance with Article 250.134 or 250.136(A) regardless of voltage of the system. Devices that are listed and identified for grounding or bonding the metallic frames of PV modules shall be permitted to bond the exposed metallic frames of PV modules to adjacent PV modules and grounded mounting structures. The panel frames are usually aluminum and care must be taken to properly make a grounding connection. A copper-bodied lug with a tin coating that is listed for direct burial must be used. These lugs are compatible with aluminum surfaces and the stainless steel hardware, and will survive the elements in an outdoor environment. These lugs should be fastened with a machine screw not a sheet metal screw, as this will better secure them. Listed modules will usually have threaded inserts in the frame for the grounding screws and should be used, see Figure 2-12 (Chapter 2). The screw and the mounting surface of the lug must be coated with an antioxidant compound rated for aluminum connections. This ensures that aluminum oxide will not form and provide a low-impedance long-lasting connection.

In accordance with Article 250.110 an equipment-grounding conductor is required between the PV array and other equipment. This means that the grounding conductor must be carried with the DC conductors and ground the enclosures, boxes, and raceways. This is further strengthened by the statement in the code that equipment-grounding conductors for the PV array shall be contained within the same raceway or cable, or otherwise run with the PV array circuit conductors when those circuit conductors leave the vicinity of the PV array.

**Size of Equipment-Grounding Conductors**—The sizing of the equipment-grounding conductors must be in accordance with Article 690.45, which states that it must follow Table 250.122. If no overcurrent protection device is used in the circuit, then the size is based on the short circuit of the PV. In the system in Figure 4-13 the overcurrent device in the combiner box protects the photovoltaic source circuit and it is rated at 10 amps. If no overcurrent protective device is used in the circuit, then the photovoltaic rated short-circuit current shall be used in accordance with Table 250.122. It is not required to increase the size of the equipment-grounding conductor for voltage drop. The equipment-grounding conductors shall be no smaller than #14 AWG. The equipment-grounding conductor in the system in Figure 4-13 would be #14 AWG copper.

**Array Equipment-Grounding Conductors**—Article 690.46 covers the array equipment conductor. As stated, if the conductor is smaller than #6 AWG, it must comply with Article 250.120(C). This requires that grounding conductors smaller than #6 AWG must be protected from damage by raceway or armor at the array.

**Grounding Electrode System**—Article 690.47 covers the requirements for the system grounding electrode(s). AC and DC systems follow the Article 250 requirements for grounding systems. The system in Figure 5-13 must comply with Article 690.47(C)(1) for systems with AC and DC grounding requirements.

1—Requires that the AC and DC grounding system be bonded together. This is done in Figure 5-13 at point 14.

2—The bonding conductor size is based on the largest of the size of the AC grounding conductor or DC grounding conductor based on Article 250.166(C), which states that the conductor size shall not be required to be larger than #6 AWG copper wire or #4 AWG aluminum wire.

**Continuity of Equipment-Grounding Systems**—Article 690.48 requires that if equipment is removed for service or repair, the system ground jumper must be installed to maintain continuity of the system ground. This requires that if the combiner box, inverter, or charge controller is removed, then a jumper must be installed while the equipment is removed.

## 2 – Photovoltaic Source Circuit

Article 690.2 defines the photovoltaic source circuit as the circuits from the photovoltaic panel to a common DC junction point and between panels. The size of the circuit conductors must be done in accordance with Articles 690.8(A-1) and (B-1). In accordance with A-1, the sum of the short circuit current of the parallel-connected modules must be multiplied by 125%. In Figure 2-16 (Chapter 2) the short circuit current is 5.61 A, therefore, the short current (Isc) is multiplied by 1.25. Isc * 1.25, so the following is 5.61 A * 1.25 = 7.01 A. But the code goes further in B-1 stating that the overcurrent devices and the conductors are sized to carry not less than 125% of the maximum currents as calculated in Article 690.8(A). Therefore the 7.01 calculated in A-1 is multiplied by 125%. 7.01 A * 1.25 = 8.76 A.

### Conductors Size

To determine the correct wiring size for the photovoltaic source circuit we must start with the current and voltage. In Figure 5-13 the corrected current is 9 A and the voltage at the photovoltaic source DC voltage is 51.6 volts DC. Then the ambient temperature, conduct fill, and the voltage drop must be factored.

A factor that must be dealt with is ambient temperature when the conductors are run with the array temperatures 70°C or higher. This means that the wiring, and any nonmetallic raceway if used, must have a minimum of 90°C rating. For example, the 9 amps in Figure 2-25. The #14 AWG THWN-2 has a current capacity of 25 A on Table 310.16 for 70°C. After a de-rating factor of 0.58 (from 310.16) this would be 14.5 A, which will work for the 9-amp load.

Voltage drop should always be reviewed at this point. For example, if the array is 200 feet from the PV to the combiner box and is carrying 9 amps at 51.6 VDC and the wire size is #14 AWG. $V_d = 2 \times K \times L \times I/CMIL$; we know from Chapter 1 that K = 12 for copper wire, L is the one-way length of the cable run 200 feet, and I = adjusted short circuit current 9 A. The voltage drop $V_d$ is calculated as:

$$V_d = \frac{2 \times 12 \times 200 \times 9}{4110}$$

$$V_d = \frac{43,200}{4110}$$

$$V_d = 10.51 \text{ volts}$$

$$V_d\% = \frac{V_d}{V}$$

$$V_d\% = \frac{10.51}{516}$$

$$V_d\% = 2.04\%$$

This would be too high for this segment of the circuit; it would be better to increase the size to #12AWG. The savings in kWs will more than pay for the increase in wire costs.

The wiring methods must meet the requirements of Article 690.31(A) This rule of the code allows for the use of all raceways and cable wiring methods in the code, and those other wiring systems and fittings that are listed and labeled for use in photovoltaic arrays shall be permitted.

### Wiring Methods

Article 690.43 requires that the equipment-grounding conductors for the photovoltaic array shall run with the photovoltaic circuit, either contained in the raceway or cable, or run with the conductors when they leave the vicinity of the array.

Article 690.31(A) states that where photovoltaic source and output circuits have maximum operating system voltages greater than 30 volts installed in readily accessible locations, circuit conductors shall be installed in a raceway.

## 2,3 – Combiner Box

The combiner box allows for the convenient termination of the photovoltaic source circuits. It contains the fuses for the photovoltaic circuit and provides the disconnect means for the photovoltaic circuits. See Figures 5-13 and 2-26 (Chapter 2).

### Disconnecting Requirements

NEC 690.13-18 sets forth the requirements for the disconnect means that are required for photovoltaic systems.

NEC 690.13 requires that there shall be a means to disconnect all current-carrying conductors of the photovoltaic circuit from all other conductors in a building or other structure. It also states that a switch, circuit breaker, or other disconnect device, either AC or DC, be installed in the ungrounded conductor if the operation of the device leaves the marked, grounded conductor in an ungrounded and energized state. The grounded conductor may have a terminal or bolted connector that allows qualified personnel to perform maintenance or troubleshooting of the equipment. This is very important as this can cause physical injury and damage to the photovoltaic equipment.

There is an exception to the requirements of Article 690.13. If the switch or circuit breaker is part of the ground fault detection system required by Article 690.5, it shall be permitted to open the grounded conductor of the photovoltaic circuit when the device is automatically opened as a normal function of the device in responding to ground faults. The device shall indicate the presence of a ground fault.

### Disconnect Marking

NEC 690.14 (C-2) requires that each disconnecting means in a photovoltaic system shall be permanently marked to identify it as a photovoltaic disconnect.

NEC 690.17 requires that where the disconnecting means may be energized in the open position, a sign shall be mounted on or adjacent to the

**Safety First**

Care must be taken when working with photovoltaic source circuits. Never disconnect the ungrounded circuit under load and never remove the fuse under load. This will cause DC arcing, which can cause physical injury and damage the fuse holder and circuit bus. It is best to cover the array and disconnect the circuit from the load whenever working on the system.

disconnecting means. The sign shall be clearly legible and have the following words or equivalent:

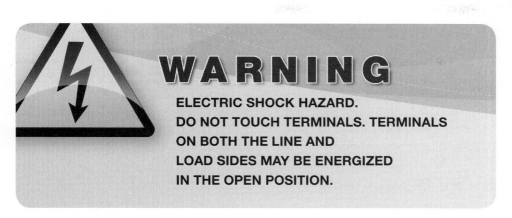

### Maximum Number of Disconnects

NEC 690.14(C-4) states that the maximum number of disconnecting means in a photovoltaic system not exceed six switches or circuit breakers. This does not mean that you cannot have more the six switches, fuses, or circuit breakers in a system—it means that you must provide a single disconnecting means in the circuit to disconnect a multi-fuse combiner box. This is not required in the system in Figure 5-13, this is done in the combiner box (4).

### Fuses

Fuses in the combiner box are energized from both sides when the circuit is energized. This is due to that fact that the photovoltaic source circuits are connected in parallel and the fuses are fed from the bus and the circuit is then connected. Article 690.16 requires that a disconnecting means be provided to disconnect a fuse from all sources of the supply if the fuses are accessible to other than qualified persons. Such a fuse in the photovoltaic source circuit shall be disconnected independently of fuses in other circuits. In the SMA combiner boxes shown in Figure 4-13 there are finger-safe fuses that disconnect the circuit without allowing the personnel to touch the fuse/fuse holder. Also there is a separate fuse for each photovoltaic circuit.

### Fuse Sizing

In an electrical system, fuses are used to protect the wiring and equipment from excessive currents. If these currents are allowed they will cause damage, heating, or in the extreme can cause fire. If the fuse is too small it could open during normal operation. If the fuse is too large it cannot provide the required protection.

The minimum sizes of the fuses are calculated using the short circuit current (Isc) of the photovoltaic source circuit. The NEC 690.8(A and B) requires that all fuses be sized for a minimum of 1.56 times the Isc of the circuit. In the photovoltaic source circuit in Figure 4-13 the Isc = 5.61 Adc, then we would determine the fuse size by Fs = 5.61 * 1.56, Fs = 8.75. The next standard fuse size would be 10A 600 VDC fuse.

## 4 – Combiner Box / Photovoltaic Output Circuit

### *Wire Size*

The photovoltaic source circuits in the combiner box are connected in parallel so the short circuit (Isc) is the sum of the Isc of the individual circuits. In the system in Figure 5-13 the $Isc = Isc_1 + Isc_2 + Isc_3 + Isc_4 + Isc_5 + Isc_6 + Isc_7 + Isc_8 + Isc_9 + Isc_{10}$, then $Isc = 5.61 * 10$ or $Isc = 56.1$ Adc. As required by 690.8 (A-B) the Isc must be multiplied by 1.56, then $Isc = 56.1 * 1.56$ or $Isc = 87.51$ Adc. The wire size must be determined using Table 310.16. The wire size must be adjusted in the field for conditions, such as the mounting location of the charge controller/inverter. If the conductors are exposed to sunlight then they must be de-rated. Also conduit fill and voltage drop must be considered.

## 5 – PV Disconnect

The NEC 2011 requires that a photovoltaic disconnecting means shall be installed at an accessible location on the outside of a building or inside the nearest entry point of the PV conductors, Article 690.14(C)(1).

NEC 690.14 (C-2) requires that each disconnecting means in a photovoltaic system shall be permanently marked to identify it as a photovoltaic disconnect.

NEC 690.17 requires that where the disconnecting means may be energized in the open position, a sign shall be mounted on or adjacent to the disconnecting means. The sign shall be clearly legible and have the following words or equivalent:

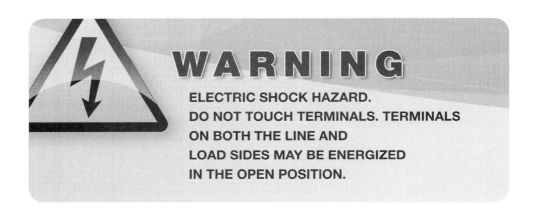

**WARNING**

ELECTRIC SHOCK HAZARD.
DO NOT TOUCH TERMINALS. TERMINALS
ON BOTH THE LINE AND
LOAD SIDES MAY BE ENERGIZED
IN THE OPEN POSITION.

## 6 – Charge Controller

There are two types of charge controllers: the first is the pulse width modulation (PWM). The PWM controller switches the power to the batteries as they charge thereby controlling the power slowly lowering the power supplied to the batteries as they get closer to charged. This type of controller allows the batteries to be more fully changed and puts less stress on the batteries. It also keeps a float charge on the batteries keeping them full charged indefinitely.

The maximum power point tracking (MPPT) is the most recent and best type of solar charge controller. The MPPT is a digital device that converts the DC into high frequency AC then looks at the output from the panels and compares it to the voltage of the batteries. It then determines the best amps to provide to the batteries and adjusts for optimal power for charging the batteries and converts it back to DC and supplies the power to the batteries. Most MPPT have efficiency between 93–97% in the conversion. The installation of the charge controller must comply with Article 690.72.

## 7 – Disconnect/Battery

This disconnect is not a NEC requirement. Article 480.5 states that a battery system over 50 volts must have a disconnecting means. I added this disconnect for maintenance of the system. If installed it should be marked as stated above.

## 8 – DC Conductors

The inverter input circuit must comply with NEC Article 690.4(B)(1). Which states that the DC conductors of the inverter input circuit shall be identified at all point of splices, termination, and connection.

## 9 – Inverter

For complete details on the inverter refer to Chapter 1. The inverter used in the example, Figure 5-13, has been designed with an inverter, which provides the required ground fault protection (GFP) and anti-islanding.

### Ground Jumper

NEC 690.49 requires that if the inverter in an interactive utility connected system is removed for maintenance, a bonding jumper shall be installed to maintain the system grounding while the inverter is removed. Bonding jumpers must be done in accordance with Article 250.120(C).

### Wire Size

NEC 690.8 requires that the maximum continuous output current of the inverter shall be used. The specification of the inverter must be consulted for the value. In the system in Figure 5-13 the maximum current ($I_m$) is 29 Aac (see specifications in Appendix Figure A-5). As required by Article 690.8(A–B) the $I_m$ must be multiplied by 1.56, then $I_m = 29 * 1.56$ or $I_m = 45.24$ Aac. The wire size must be determined using Table 310.16. The wire size must be adjusted in the field for conditions, such as the mounting location of the inverter. If the conductors are exposed to sunlight, then they must be de-rated. Also conduit fill and voltage drop must be considered.

## 10 – Disconnect Means

NEC 690.14(C) states that means shall be provided to disconnect all conductors in a building or other structure from the photovoltaic system conductors.

In accordance with NEC 690.15, means shall be provided to disconnect the inverter(s) from all ungrounded conductors of all sources.

NEC 690.17 requires that where the disconnecting means may be energized in the open position, a sign shall be mounted on or adjacent to the disconnecting means. The sign shall be clearly legible and have the following words or equivalent:

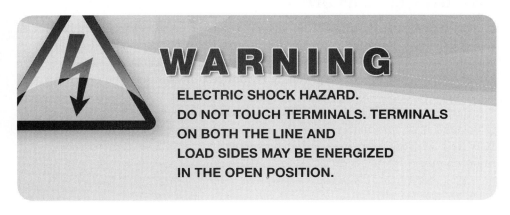

**WARNING**

ELECTRIC SHOCK HAZARD.
DO NOT TOUCH TERMINALS. TERMINALS
ON BOTH THE LINE AND
LOAD SIDES MAY BE ENERGIZED
IN THE OPEN POSITION.

## 12 – Circuit breaker

The circuit breaker is the point of connection for the interactive inverter system. The rules of Article 690.45 must be followed for the connection to the main panel. In the system in Figure 5-13 the connection is made to the load side of the service entrance equipment and must follow the rules in Articles 690.45(B1) through (B7).

1—It is required that each inverter or interconnecting system must have a dedicated fuse or circuit breaker.

2—The sum of the current rating of the overcurrent devices shall not exceed 120% of the rating of the bus bar or conductor.

3—The connection point shall be on the line side of the GPF equipment.

4—Equipment containing multiple devices with multiple sources shall be marked to indicate the presence of all sources.

5—If the circuit breakers are back fed, then the circuit breaker shall be suitable for such operation. If a circuit breaker is marked line and load, then it has been tested only in that direction.

6—Listed plug-in-type, back fed circuit breakers from utility-interactive inverters complying with Article 690.60 shall be permitted not to have the additional fastener normally required by Article 480.36(D). This is allowed because if the breakers are lifted from the bus bar, then the inverter would trigger the anti-islanding function and shut off the AC output immediately as per Article 690.61.

7—The circuit breakers feeding the panel must be at the opposite end of the bus bar (load end) from the main breaker (line end) and be permanently marked with the following or equivalent.

**WARNING**

INVERTER OUTPUT CONNECTION
DO NOT RELOCATE THIS
OVERCURRENT DEVICE

## 13 – Storage Batteries

Battery systems in a dwelling or home can be dangerous. Read and comply with all relevant articles of the NEC 2011. Safety is very important when dealing with high currents. An object dropped on a battery terminal can cause an explosion. The safety of the occupants of the dwelling is your responsibility. So be safe and guard the batteries well.

Batteries in a solar system are subject to many charge and discharge cycles. If flooded lead acid batteries are used this will require frequent maintenance. It is best therefore to use sealed batteries in a small system to reduce the maintenance.

The batteries are installed on a rack system, which must comply with NEC Article 480.8 and must be grounded as defined in Article 250.110.

The installation of the battery system shall comply with all articles in 690.71. Article 690.71(B)(1) states that storage batteries in a dwelling shall operate at less than 50 volts nominally. The maximum number of cells allowed in a dwelling is twenty-four 2-volt cells connected in series.

Article 690.71(B2) covers guarding of live parts. This is very important as to prevent persons or objects from touching live terminals.

## REVIEW

1. Define the basic battery.
2. Is an anode a positive or negative electrode?
3. A secondary cell battery cannot be charged.
   A. True
   B. False
4. What is the specific gravity of water?
   A. 1.00
   B. 1.84
   C. 0.01
5. What is the hydrometer used for?
6. Three batteries in parallel each have a current rating of 50 amps. What is the total current?
7. Three batteries in series each have a current of 50 amps. What is the total current?
8. With two 12-volt batteries in a series circuit what is the total voltage?

# CHAPTER 6

## HYDROGEN FUEL CELL

### objective

This chapter will expose the reader to the electrical systems that support the hydrogen fuel cell systems. By the end of this chapter the reader will have a basic understanding of hydrogen fuel cells and be able to work on these systems with an understanding of the system and relevant National Electrical Code.

## WHAT WE NEED TO KNOW

We tend to think of the hydrogen fuel cell as new technology. But point of fact, the principle of the hydrogen fuel cell was discovered by German scientist Christian Friedrich Schonbein in 1838.

## INTRODUCTION

Figure 6-1 is my attempt to redraw the first fuel cell demonstrated by Sir William Robert Grove in 1839 and later sketched in 1842. Grove's work was based on the work of Schonbein. When Sir William Grove disconnected the battery from his electrolyzer and connected the leads together, he observed a current flowing in the opposite direction consuming the hydrogen and oxygen gases. He called his device the "gas battery." The fuel cell that he made used similar materials to today's phosphoric-acid fuel cell. His device was constructed with platinum electrodes placed in test tubes of hydrogen and oxygen in a bath of dilute sulfuric acid. It generated a voltage of just under a volt. Due to problems of corrosion of the electrodes and instability of the materials, his fuel was not practical. Little research or development of fuel cells was done for many years to follow.

Significant work on the fuel cells did not resume until the 1930s when a chemical engineer at Cambridge University, by the name of Francis Bacon, began serious work on the fuel cell again.

In 1955, W. Thomas Grubb, a chemist working for General Electric Company (GE), further modified the original design by using sulphonated polystyrene ion-exchange membrane as the electrolyte. Three years later, Leonard Niedrach, another GE chemist, devised a way of depositing platinum onto the membrane, which served as a catalyst for the necessary hydrogen oxidation and oxygen reduction reactions. This became known as the "Grubb-Niedrach fuel cell," which was the first polymer electrolyte membrane (PEM) fuel cell. GE further developed this technology with NASA and McDonnell Aircraft, which lead to the use of these fuel cells in the U.S. space program's projects—the Gemini Project was the first. This was the first commercial use of a hydrogen fuel cell.

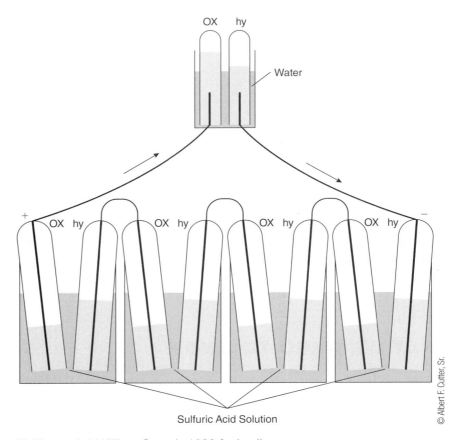

OX hy

Water

+ OX hy OX hy OX hy OX hy −

Sulfuric Acid Solution

© Albert F. Cutter, Sr.

■ **Figure 6-1** William Grove's 1839 fuel cell.

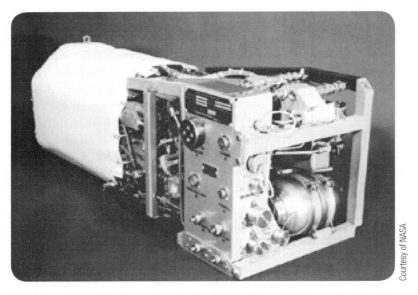

Courtesy of NASA

■ **Figure 6-2** UTC Power's alkaline fuel cell module from the Apollo spacecraft.

In the 1960s United Technologies Corporation's UTC Power subsidiary was the first company to manufacture and commercialize a large stationary fuel cell system for use as a co-generation power plant. UTC Power continues to supply fuel cells to NASA for use in space vehicles having supplied them for the Apollo missions and the Space Shuttle program (Figure 6-2). As part of the $58.7 billion United

Courtesy of MTI MicroFuel Cells, Inc.

**Figure 6-3** Handheld hydrogen fuel cell.

Technologies Corp., UTC Power has the resources, stability, and commitment to help redefine the energy practices of a generation. They are developing fuel cells for automobiles, buses, and cell phone towers; the company has demonstrated the first fuel cell capable of starting under freezing conditions with its proton exchange membrane (PEM). UTC Power remains the world leader in fuel cell technology.

Thanks to a strong commitment of the United States government and other countries around the world, research continues today on the use of fuel cells for everything from smartphones to cities. Our use of hydrogen fuel cells is expanding at a rapid rate; knowledge of these systems is important for our near future.

Shown in Figure 6-3 is the MTI Micro Mobion, which is a handheld hydrogen fuel cell power supply that can recharge a smartphone about seven times. This generator uses a direct methanol fuel cell (DMFC), which combines methanol with oxygen and water from the air to generate electricity.

Angstrom Power Inc., a Vancouver, Canada–based company, together with Motorola US, a mobile phone manufacture, have completed a six-month trial of Angstrom Power's new technology that will reinvent the way that cell phones are powered. Their revolutionary new fuel cell will replace the lithium batteries currently used today.

The "micro hydrogen" platform uses advanced micro-fluidics and innovative fuel cell architecture and a refillable hydrogen storage tank. Angstrom and Motorola used the MOTOSLVR L7 mobile phone to implement the new platform, without changing the look or size of the phone. The phone managed to run twice the talk time of the lithium battery with the hydrogen-powered platform.

# OVERVIEW

In my introduction to this manuscript, I stated that man has been in a continual search for a source of renewable non-polluting alternative energy. I feel that the hydrogen fuel cell is the energy source of the future. Unlike fossil fuels and conventional batteries, it does not pollute the environment. Hydrogen fuel cells produce pure water and useable heat as a byproduct. In combination solar, wind, and hydroelectric plants can produce a never ending source of clean electrical power.

Courtesy of NASA

■ **Figure 6-4** Methanol fuel cell.

Hydrogen is a versatile energy carrier that can be used to power every device that an end user can use. From handheld devices to entire cities—even with fuel sources like methanol (Figure 6-4), which produce small amounts of carbine dioxide—fuel cells are the cleanest, most efficient source of electrical power production.

The fuel cell is an electrochemical energy conversion device that efficiently captures and uses the power of hydrogen.

Stationary fuel cells can be used to power remote locations, as a source of backup power, and for distributed power generation and cogeneration power plants that use the excess heat for other applications. Fuel cells can be used in any application that we currently use batteries.

Fuel cells will also power our transportation needs—anything that can be driven by electricity can use fuel cells. This will greatly reduce our dependency on expensive, polluting fossil fuels.

# FUEL EFFICIENCIES

A conventional combustion-based power plant generates electricity at a 33 to 35% efficiency rate. Fuel cells generate electricity at efficiencies of 40% and 80% with cogeneration systems that use the excess heat.

Under normal driving conditions a conventional car with a gasoline engine is less the 20% efficient, while a hydrogen fuel cell vehicle with electric motors uses 40 to 60% of the fuel's energy. In addition fuel cells are quiet and have few moving parts which make them well suited for many applications.

| Fuel Cell Type | Operating Temperature | System Output | Efficiency | Applications |
|---|---|---|---|---|
| Alkaline (AFC) | 90–100°C 194–212T | 10 kW–100 kW | 60–70% electric | • Military<br>• Space |
| Phosphoric Acid (PAFC) | 150–200°C 302–392°F | 50 kW–1 MW (250 kW module typical) | 80–85% overall with combined heat and power (CHP) (36–42% electric) | • Distributed generation |
| Polymer Electrolyte Membrane or Proton Exchange Membrane (PEM)* | 50–100°C 122–212°F | <250 kW | 50–60% electric | • Back-up power<br>• Portable power<br>• aSmall distributed generation<br>• Transportation |
| Molten Carbonate (MCFC) | 600–700°C 1112–1292T | <1 MW (250kW module typical) | 85% overall with CHP (60% electric) | • Electric utility<br>• Large distributed generation |
| Solid Oxide (SOFC) | 650–1000°C 1202–1832T | 5 kW–3 MW | 85% overall with CHP (60% electric) | • cAuxiliary power<br>• Electric utility<br>• Large distributed generation |

Source: Argonne National Laboratory

*Direct Methanol Fuel Cells (DMFC) are a subset of PEMFCs typically used for small portable power applications with a size range of about a subwatt to 10 MW and operating at 60–90°C.

**Figure 6-5** Fuel cell chart.

Courtesy of DOE

## COMPARISON OF FUEL CELL TECHNOLOGIES

Do not get confused by the chart shown in Figure 6-5. Fuel cells are similar to conventional batteries in many ways, but contrast in other ways. They have an electrolyte and two electrodes: a cathode and an anode. This basic configuration is used by all fuel cells. There are many types of fuel cells, classified by the type of electrolyte used, which is similar to conventional batteries, which have alkaline, lithium, etc. As shown in Figure 6-5, the left-hand column is the type of electrolyte used by the fuel cell; the rest of the chart is self-explanatory. I draw your attention to the efficiency ratings, as you can see these are very impressive figures. With this efficiency you cannot help ask why? Why would you have large fossil fuel burning generators for backup power when you can have a quiet non-polluting fuel cell? I have worked on computer data centers where they had 14 large generators that when the power failed and they kicked in, you could hear them for miles. This is to say nothing about all the fuel that they used to run for short periods.

## HOW DO FUEL CELLS WORK

First we have to declare the meaning of the term "Fuel Cell." The term fuel cell can be used for a single cell or a complete stack of cells connected in series or a complete generator system that can have many stacks in series/parallel configuration. A single fuel cell consists of an electrolyte sandwiched between two electrodes, an anode, and a cathode. Bipolar plates on either side of the cell help to distribute gasses and serve as collectors, as shown in Figure 6-6. They are much like batteries except that batteries have one solid metal electrode that is slowly consumed as the electricity is produced. In a fuel cell the electrodes are not consumed, and the fuel cell can produce electricity as long as more fuel (hydrogen) and an oxidizer (oxygen or air) are pumped through it. In a Polymer electrolyte membrane (PEM) fuel cell hydrogen gas flows through channels to the anode, where a catalyst causes the hydrogen molecules to separate into protons and electrons. While protons are conducted through the membrane to the other side of the cell, the stream of negatively charged electrons follows an external circuit to the cathode.

On the oxygen side of the cell, oxygen, which is drawn from the air, flows through channels to the cathode. When the electrons return from the external circuit they react with the oxygen and the hydrogen protons, which have moved through the membrane, at the cathode to form water. This is an exothermic reaction, which generates heat that can be used outside the fuel cell.

A hydrogen fuel cell is a very efficient source of DC current, from an electrician's point of view it is just a more efficient battery and it is used in the following circuits that way.

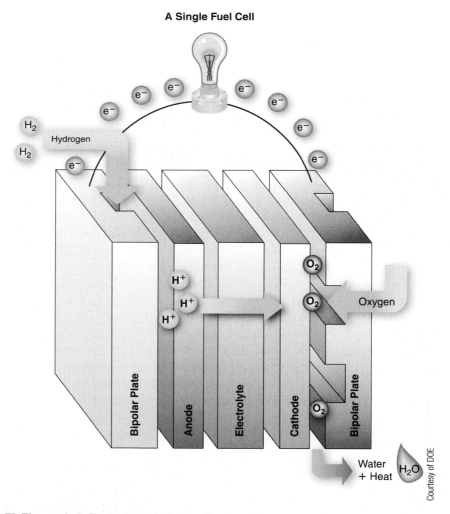

**A Single Fuel Cell**

Courtesy of DOE

■ **Figure 6-6** Typical single fuel cell.

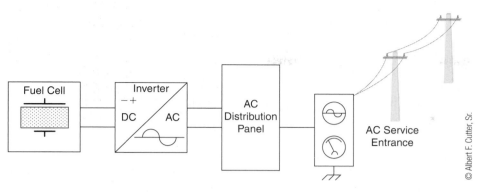

■ **Figure 6-7** Fuel cell generator system.

# FUEL CELL CIRCUITS

As shown in Figure 6-7, the basic fuel cell generator is a source of DC power, which is fed directly into an inverter to supply AC power to the load and grid. This is a basic design for the system.

## Single-Phase System

Before any fuel cell system is installed you must consult all federal, state, and local codes.

Article 692 of the National Electrical Code (NEC) covers the use of fuel cells systems.

As with any complex electrical system only qualified and trained electricians should install the system and all associated wiring.

Article 692.4(C) requires that fuel cell system installation, including all associated wiring and equipment connections, shall be done by qualified persons.

Article 692.6 requires that the fuel cell shall be evaluated and listed prior to installation.

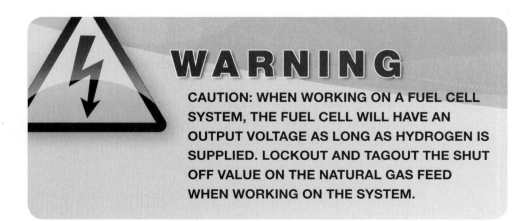

**WARNING**

CAUTION: WHEN WORKING ON A FUEL CELL SYSTEM, THE FUEL CELL WILL HAVE AN OUTPUT VOLTAGE AS LONG AS HYDROGEN IS SUPPLIED. LOCKOUT AND TAGOUT THE SHUT OFF VALUE ON THE NATURAL GAS FEED WHEN WORKING ON THE SYSTEM.

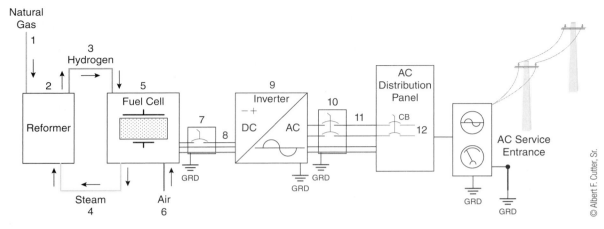

**Figure 6-8** 120/240 VAC fuel cell system.

© Albert F. Cutter, Sr.

Use the numbered items on figure 6-8 for the following discussion:

1. Natural gas feed – If the system is required to be shut off for any reason, the main value on the natural gas feed must be locked out to prevent risk of electrical shock or arching. Article 692.54 states that the location of the manual fuel shut-off value shall be marked at the primary disconnecting means of the circuits supplied or building (Figure 6-8, #10).

2. Reformer – Fossil fuel reforming is a method of producing hydrogen from fossil fuels—in this case natural gas. The reformer reacts steam at high temperature with the fossil fuel and produces hydrogen.

3. The output of the reformer is hydrogen which is filtered to remove any ammonia before it enters the fuel cell.

4. Steam is output from the fuel cell to be used by the reformer. The steam is run through a heat exchanger to provide useable heat from the fuel cell. This can be connected to the building's hot water system to heat or cool the building.

5. Fuel cells can be any type of electrolyte fuel cell stack or group of stacks. Most commercial stationary systems are alkaline (AFC), phosphoric acid (PAFC), and polymer electrolyte membrane or proton exchange membrane (PEM).

6. Oxygen (air) is fed into the fuel cell.

7. DC disconnect is required for maintenance and Article 692.13 requires that a disconnect means is provided on all current-carrying conductors of the fuel cell power system.

8. DC circuit conductors – The fuel cell nameplate rating shall be used for circuit sizing, per Article 692.8. Voltage drop must be taken into account.

9. The inverter must be rated for the maximum continuous output current of the fuel cell.

In an interactive system the inverter must provide for anti-islanding to protect the grid from being energized in a condition when the grid power is interrupted, per Article 692.62. The inverter must remain off-line for a time, usually 5 minutes, when the grid power is returned to allow the grid power to stabilize.

10. Disconnect means is required for system maintenance and Article 692.13 requires that the system shall have a disconnect means of all current-carrying conductors from the building system conductors.

Article 692.17 states that this can be a manually-operated switch or circuit breaker and where all terminals of the device can be energized in the off state shall be marked as follows:

Article 692.53 also requires that the fuel cell disconnect shall be marked with the following:

- Output voltage
- Output power rating
- Continuous output current rating

11. Inverter output circuits are sized according to the nameplate rating, Article 692.8.

12. The circuit breaker, which is the point of connection must comply with Article 705.12.

**⚡ DANGER**

Electric shock hazard. Do not touch terminals. Terminals on both the line and load sides may be energized in the open position.

## Three-Phase System

Use Figure 6-9 for the following:

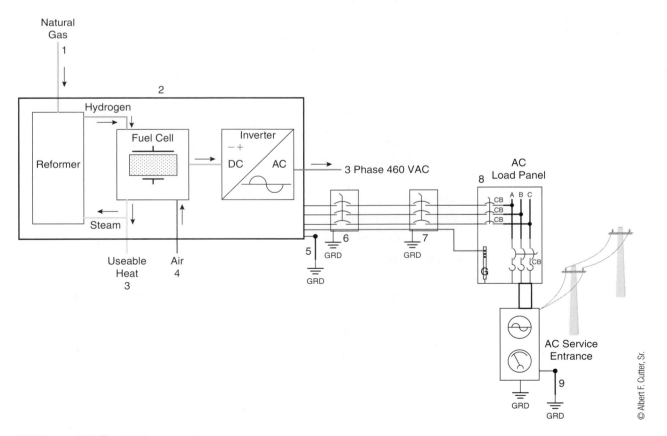

**Figure 6-9** Three-phase system.

Use the numbered items on figure 6-9 for the following discussion:

1. Natural Gas Feed – This could also be LP, methanol, or other fossil fuels depending on the design of the fuel cell. Always read the specification of the system to familiarize yourself with the entire system. Article 692.54 states that the location of the manual fuel shut-off value shall be marked at the primary disconnecting means of the circuit or building (Figure 6-9 Item 7). This is required in case the system must be shut down. The fuel cell will continue to generate power until the value is closed.

2. Stationary Fuel Cell System – The majority of stationary fuel cell systems made today are self-contained units. Article 692.6 requires that the fuel cell shall be evaluated and listed prior to installation. Always read the operations manual and specifications to familiarize yourself with the complete system.

3. Useable Heat – The fuel cell generates heat during the normal operation of the system. This heat can be and should be used to heat or cool the building.

4. Air input – Oxygen is required for the fuel cell reaction.

5. Ethernet Cable – Units like the UTC PowerCare 400 fuel cell have a microprocessor that manages and monitors the system. This processor can be linked to the building management computer system. The CAT5 cable must be in a separate conduit.

6. Ground Electrode – The manufacturer will require a grounding electrode for the system. Article 692.47 states that this shall be connected to the equipment-grounding conductor.

7. Maintenance Disconnect – Article 692.13 requires that a disconnect means is provided on all current-carrying conductors of the fuel cell power system. This also assists with any maintenance required on the system.

**DANGER**

Electric shock hazard. Do not touch terminals. Terminals on both the line and load sides may be energized in the open position

Article 692.17 states that this can be a manually-operated switch or circuit breaker, and where all terminals of the device can be energized in the off state shall be marked as follows.

Article 692.53 also requires that the fuel cell disconnect shall be marked with the following:

- Output voltage
- Output power rating
- Continuous output current rating

8. Disconnect – Article 692.13 requires that the system shall have a disconnect means for all current-carrying conductors from the building system conductors. This disconnect must be marked as stated above in 7.

9. Building loadcenter

10. Service Ground Electrode

The ClearEdge5 is a 5 kW stationary co-generation (combined heat and power CHP) hydrogen fuel system (Figures 6-10 and 6-11). ClearEdge5 converts natural gas

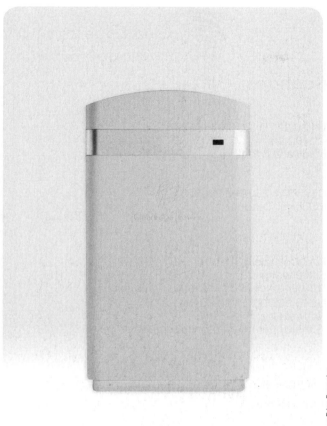

Courtesy of ClearEdge Power Inc.

**Figure 6-10** ClearEdge5 5 kW fuel cell system.

to produce ultra-clean hydrogen, which is converted by the PEM fuel stack to generate AC power and useable heat. The ClearEdge5 is a complete self-contained system and complies with ISO 9001 and ISO 14001 standards. Using the heat and power from the system it achieves an efficiency rating reaching 90%. ClearEdge Power was established in 2003 and has been supplying systems up and down the West Coast.

UTC Power has more than 270 stationary fuel cell systems installed worldwide. The system in Figure 6-12 is a UTC Power PureCell system installed at Price Chopper in Colonie, New York.

With the release of the UTC Power 400 kW PureCell Model 400 we now have a highly efficient source of clean energy for large-scale stationary installations.

Shown in Figure 6-13 is the specification for the PureCell Model 400. It is a 400 kW three-phase system that has a lifetime of 10 years, with scheduled maintenance in 5 years. When you look at the efficiency rating of 42% for the fuel cell electrical conversion and an overall rating of 90% when used as a co-generation system, this is a very practical energy source. The emissions are minimal compared to fossil fuel generators. The system is a closed loop system; it produces its own

# ClearEdge5 Product Specifications

## Equipment Product Specifications

| | |
|---|---|
| Dimensions | W: 36.02" D: 26.32" H: 67.75" |
| Weight | ~1,400 lbs |
| Noise | 60 dBa @ 3 feet |
| Seismic Design Category | "E" |
| Site Class | "D" |
| Certifications and Registrations | CSA/FC1-2004/NFPA 853 |
| Location | Indoor or Outdoor |

## System Summary

| | |
|---|---|
| Energy System | Fuel Cell microCHP |
| Rated Power | 5,000 watts |
| | (120/240V or 208V AC/grid compatible) |
| Heat Output | Up to 20,000 BTU/hour at 150° F |
| Fuel | Natural Gas |
| Feed Stock | ½ therm per hour (12 therms per day) |

## Project Site Infrastructure Requirements

1. 120/240VAC/60Hz 40 amp for start-up of the unit
2. Refer to system drawings for 208VAC requirements
3. Disconnects per local code
4. CAT 5 network connection (high-speed Internet) for remote monitoring service
5. Natural gas line (minimum ¾" at 7-14" wc input at 50k BTUs per hour)
6. 1,000cfm ventilation (for indoor installations)
7. All ClearEdge5 vents must be kept clear at all times
8. All hydronic lines minimum ¾" insulated to R-5
9. Hydronic lines maximum 25' head pressure
10. 2 GPM minimum water flow for hydronics
11. Backflow device regulator set at a minimum pressure of 15 psi and minimum flow 2 GPM
12. Space accommodations for 28" diameter hot water storage tank (if required)

Page 1 of 3

■ **Figure 6-11** Specifications for ClearEdge5 fuel cell system.

Courtesy of UTC Power Corporation

■ **Figure 6-12** UTC PureCell installation.

water as you will see in Figure 6-14. It is a very quiet system with a noise rating of less than <60 db at 33 feet. As we move from a central grid-based electrical system to a distributed electrical system in the near future, systems like these from UTC Power will play a major role in our becoming fossil fuel independent.

Figure 6-14 is an inside view of the PureCell system. Natural gas is fed into the reformer where it is filtered and processed along with the steam produced by the fuel cell. This process generates the hydrogen that the fuel cell uses. The fuel cell produces steam that is fed to the reformer and usable heat is produced as a byproduct. This heat can be used to heat and cool the facility. Air is fed into the fuel cell and DC power is produced. This DC power is then fed into the onboard inverters and the output of the system is useable 400 kW of 3 AC phase power.

# MODEL 400
## PureCell® System

**Introducing a new generation of fuel cell technology:**

**The PureCell® Model 400 Energy Solution.**

UTC Power is a world leader in developing and producing fuel cells for on-site power, transportation, space and defense applications. We are committed to providing high quality solutions for the distributed energy market that increase energy productivity, energy reliability and operational savings for our customers. Building on our unmatched operational experience and a technology platform proven at more than 260 sites worldwide, UTC Power is pleased to offer an advanced fuel cell energy solution for the commercial marketplace.

The ultra clean and quiet PureCell® Model 400 fuel cell can provide up to 400 kW of assured electrical power, plus up to 1.7 million Btu/hour of heat, for combined heat and power applications. And with energy efficiencies more than double those of traditional power sources, the PureCell® Model 400 system is an energy solution that will not only help you conserve precious resources, it will save you money, shield you from operational interruption, and secure your place at the forefront of environmentally sustainable business practices.

## Performance Characteristics

### Power

| | |
|---|---|
| Electric power | 400 kW/400 to 471 kVA initial |
| | 400 kW lifetime average |
| | 360 kW initial (ADG) |
| Voltage/frequency | 480VAC/60 Hz/3 phase** |
| | 400VAC/50 or 60 Hz/3 phase |

### Efficiency

| | |
|---|---|
| Electrical (LHV) | 42% initial/40% nominal (5 yr) |
| Overall (LHV) | 90%*** |

### Fuel

| | |
|---|---|
| Supply | Natural gas or ADG† |
| Consumption (HHV) | 3.60 MMBtu/hr (1,054 kW) initial |
| | 3.79 MMBtu/hr (1,110 kW) average |
| | 3,493 scfh (98.9 Nm3/hr) initial |
| | 3,678 scfh (104.2 Nm3/hr) average |
| Pressure | 4 to 14 in. water (1.0 to 3.5 kPa)‡ |

### Heat Recovery

| | |
|---|---|
| Low grade (140°F/60°C supply)§ | 1.537 MMBtu/hr (450 kW) initial |
| | 1.708 MMBtu/hr (500 kW) nominal |
| High grade (250°F/121°C supply)§ | 0.683 MMBtu/hr (200 kW) initial |
| | 0.785 MMBtu/hr (230 kW) nominal |

### Emissions*

| | |
|---|---|
| NOx | 0.035 lb/MWh (0.016 kg/MWh) |
| CO | 0.008 lb/MWh (0.004 kg/MWh) |
| CO2 | 1120 lb/MWh (508 kg/MWh) average |
| SOx | Negligible |
| Particulate matter/VOCs | Negligible |

### Water

| | |
|---|---|
| Consumption | None (up to 86°F/30°C ambient) |
| Discharge | None (normal operating conditions) |

### Other

| | |
|---|---|
| Noise | <65 dBA at 33 ft (10m) with no heat recovery |
| | <60 dBA at 33 ft (10m) with full heat recovery |
| Overhaul interval | 10 yr |

*Emissions meet 2007 California Air Resources Board standards. **Operating range from -20° to 113°F (-29° to 45°C) at up to 492 ft (150m). *** Overall efficiency as given assumes full thermal utilization. † All given characteristics are for a natural gas application, unless otherwise noted. Maximum allowable levels for natural gas components are documented separately in the PureCell® Model 400 Data and Applications Guide. ADG applications require an additional gas processing unit. ‡ Gauge pressure. § Available heat at rated power. Low-grade heat assumes a return temperature of 80°F (27°C), high-grade heat assumes a return temperature up to 230°F (110°C). If high-grade heat is utilized, the remaining value will be available as low-grade heat.

**UTC Power**
A United Technologies Company

**MODEL 400**
PureCell® System

**Figure 6-13** UTC PureCell 400 specifications.

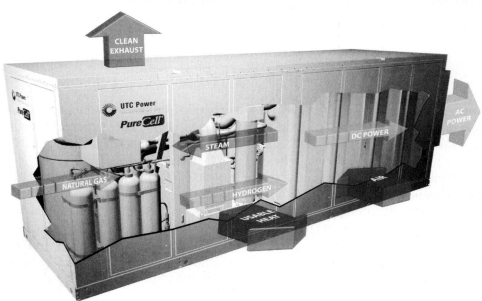

**Figure 6-14** UTC PureCell stationary phosphoric acid fuel cell (PAFC) power plant.

Shown in Figure 6-15 is a co-generation hybrid system, which, until recently, was not cost-effective. Due to recent advancements in electrolyzers made by engineer Richard Bourgeois and his colleagues at the General Electric research facility in Niskayuna, NY, these systems are now cost-effective. An electrolyzer uses electricity to free hydrogen from water; 95% of the hydrogen currently produced today is from natural gas. Producing hydrogen from electricity is a much smarter green alternative. Bourgeois's prototype replaces the costly tooled metal

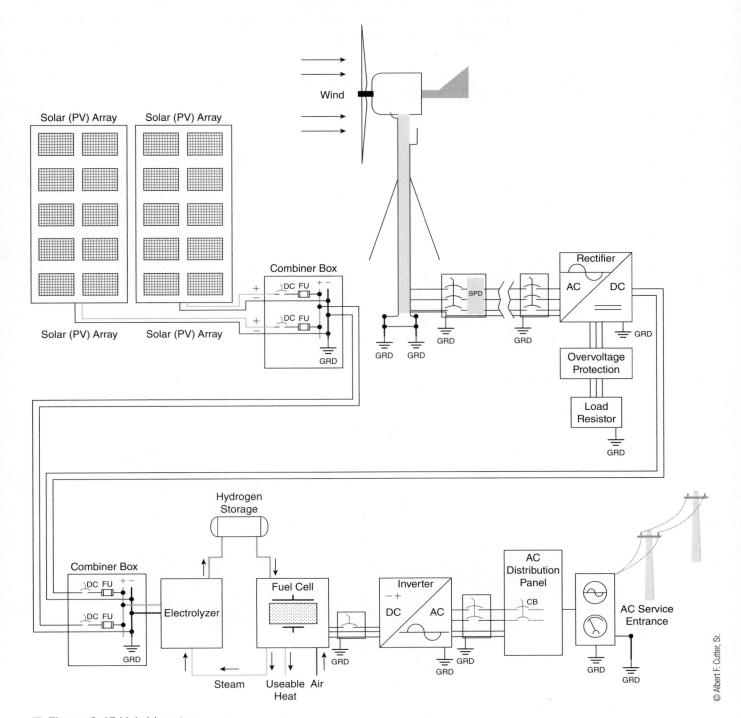

■ **Figure 6-15** Hybrid system.

parts with a moldable high-tech GE plastic called Noryl. The savings on materials, manufacturing, and assembly provides a cost saving that will allow the cost of a kilogram of hydrogen to be about $3 instead of the current $6 to $8; steam reforming of hydrogen from natural gas cost between $4 and $5. The system uses the solar array and the wind turbine to generate DC current that is then used by an electrolyzer to provide the hydrogen. The hydrogen is stored in a storage tank until needed. In this configuration the system will continue to generate useable AC power and heat when the solar array is not generating power from the sun and the wind turbine is not producing power. In the near future, systems like this will be used to produce the electricity that we need to power our homes, offices, factories, and the vehicles that we drive. Cost-effective alternative green energy is not coming tomorrow, it is here today.

## REVIEW

1. Hydrogen is a _____.
   A. Fuel
   B. Carrier
   C. Energy Source

2. The efficiency of a co-generation fuel cell system is _____.
   A. About 40%
   B. About 70%
   C. About 90%

3. Hydrogen fuel cell requires oxygen and _____ to function.

4. PEM stands for _____ _____ _____.

5. Article 692.4(C) states that anyone can install a fuel cell system.
   A. True
   B. False

6. A reformer converts air into hydrogen.
   A. True
   B. False

7. Today 95% of the hydrogen is produce from _____ _____.

8. CHP stands for _____ _____ _____.

9. An electrolyzer converts electricity to hydrogen gas.
   A. True
   B. False

10. The current cost of producing hydrogen gas from natural gas is between $6 and $12 per kilogram of hydrogen.

    A. True
    B. False

# APPENDIX: Product Specifications

Advanced Contact Technology

**Multi-Contact**

**MC**
*STÄUBLI GROUP*

| | |
|---|---|
| **Kupplungsbuchse, -stecker MC4** | **Female and male cable coupler MC4** |

Kupplungsbuchsen und -stecker als Einzelteil (inklusive Isolierteil)

Female and male cable coupler as individual part (including insulating part)

**PV-KBT4...**

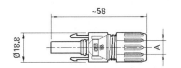

**PV-KST4...**

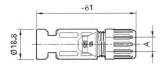

| Technische Daten | Technical data | |
|---|---|---|
| Steckverbindersystem | Connector system | Ø 4mm |
| Bemessungsspannung | Rated voltage | 1000V DC (IEC) 600V DC (UL) |
| Bemessungsstrom | Rated current | 17A (1,5mm²) 22,5A (2,5mm²; 14AWG) 30A (4mm², 6mm²; 10AWG) 43A (10mm²) |
| Prüfspannung | Test voltage | 6kV (50Hz, 1min.) |
| Umgebungstemperaturbereich | Ambient temperature range | -40°C...+90°C (IEC) -40°C...+75°C (UL) -40°C...+70°C (UL: 14AWG) |
| Obere Grenztemperatur | Upper limiting temperature | 105°C (IEC) |
| Schutzart, gesteckt ungesteckt | Degree of protection, mated unmated | IP67 IP2X |
| Überspannungskat./Verschmutzungsgrad | Overvoltage category/Pollution degree | CATIII/2 |
| Kontaktwiderstand der Steckverbinder | Contact resistance of plug connectors | 0,5mΩ |
| Schutzklasse | Safety class | II |
| Kontaktsystem | Contact system | MC Kontaktlamellen MC Multilam |
| Anschlussart | Type of termination | Crimpen Crimping |
| Kontaktmaterial | Contact material | Kupfer, verzinnt Copper, tin plated |
| Isolationsmaterial | Insulation material | PC/PA |
| Verriegelungssystem | Locking system | Snap-in |
| Flammklasse | Flame class | UL94-V0 |
| Kabelzugentlastung gemäss | Cable strain relief according to | EN 50521:2008 |

**Sicherungshülse** Seite 53
**Verschlusskappen** Seite 55
**Montageschlüsselset** Seite 61

**Safety lock clip** page 53
**Sealing caps** page 55
**Assembly tools** page 61

[MA] Montageanleitung **MA231**
*www.multi-contact.com*

[MA] Assembly Instructions **MA231**
*www.multi-contact.com*

14

*www.multi-contact.com*

Courtesy of Muli-Contact USA

■ **Figure A-1**

Advanced Contact Technology

**Multi-Contact**

*STÄUBLI GROUP*

- Snap-In Verriegelung
- Durch Einsatz der Sicherungshülse PV-SSH4 Verriegelung nach NEC 2011, nur mit Werkzeug entriegelbar
- Bewährte, langzeitstabile MC Lamellentechnologie
- Bewährter Steckverbinder
- Auch für Querschnitte von 10mm² konfektionierbar
- Auch erhältlich als konfektionierte Leitungen, siehe Seite 66
- Leitungen nach Kundenwunsch, siehe Seite 68

- Snap-in locking
- Locking by safety lock clip PV-SSH4 in accordance with NEC 2011, can be released only with tool
- Proven MC-Multilam technology with high long-term stability
- Tried and tested plug connectors
- Available for assembly with cross-sections of 10mm²
- Also available as ready made leads, see page 66
- Leads made to customer's specifications, see page 68

| Typ Type | Bestell-Nr. Order No. | Kupplungsbuchse Female cable coupler | Kupplungsstecker Male cable coupler | Ø-Bereich Kabelverschraubung Ø range of cable gland | Leiterquerschnitt Conductor cross section | | Zulassungen Approvals |
|---|---|---|---|---|---|---|---|
| | | | | A (mm) | mm² | AWG | b (mm) |
| PV-KBT4/2,5I-UR | 32.0010P0001-UR | × | | 3 – 6 | 1,5; 2,5 | 14 | 3 |
| PV-KST4/2,5I-UR | 32.0011P0001-UR | | × | 3 – 6 | 1,5; 2,5 | 14 | 3 |
| PV-KBT4/2,5II-UR | 32.0012P0001-UR | × | | 5,5 – 9 | 1,5; 2,5 | 14 | 3 |
| PV-KST4/2,5II-UR | 32.0013P0001-UR | | × | 5,5 – 9 | 1,5; 2,5 | 14 | 3 |
| PV-KBT4/6I-UR | 32.0014P0001-UR | × | | 3 – 6 | 4; 6 | 10 | 5 |
| PV-KST4/6I-UR | 32.0015P0001-UR | | × | 3 – 6 | 4; 6 | 10 | 5 |
| PV-KBT4/6II-UR | 32.0016P0001-UR | × | | 5,5 – 9 | 4; 6 | 10 | 5 |
| PV-KST4/6II-UR | 32.0017P0001-UR | | × | 5,5 – 9 | 4; 6 | 10 | 5 |
| PV-KBT4/10II | 32.0034P0001 | × | | 5,5 – 9 | 10 | – | 7,2 |
| PV-KST4/10II | 32.0035P0001 | | × | 5,5 – 9 | 10 | – | 7,2 |

Courtesy of Multi-Contact USA

*www.multi-contact.com*

**█ Figure A-2**

15

Advanced Contact Technology

**Multi-Contact**

*STÄUBLI GROUP*

## Kupplungsbuchse, -stecker MC3

**Kupplungsbuchsen und -stecker als Einzelteil**
(inklusive Isolierteil)

## Female and male cable coupler MC3

**Female and male cable coupler as individual part**
(including insulating part)

PV-KBT3...

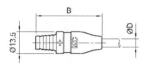

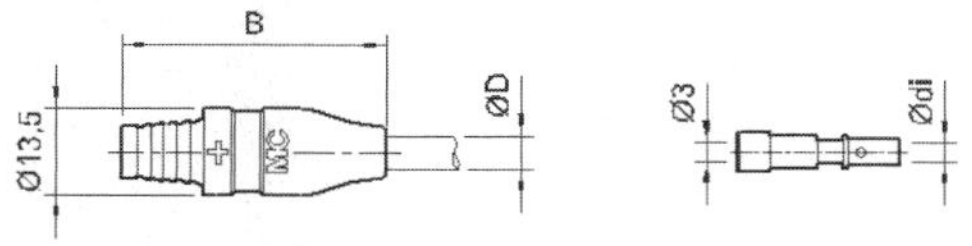

PV-KST3...

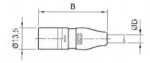

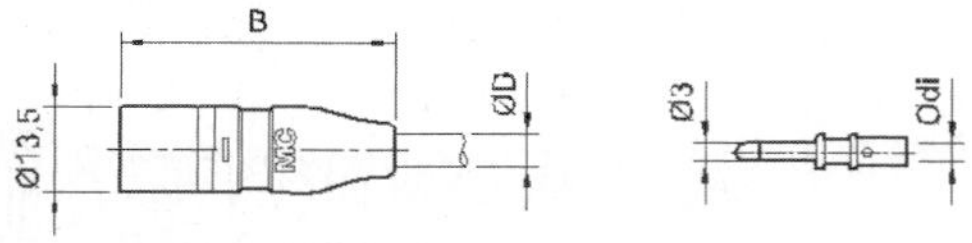

| Technische Daten | Technical data | |
|---|---|---|
| Steckverbindersystem | Connector system | Ø 3mm |
| Bemessungsspannung | Rated voltage | 1000V DC (IEC)<br>600V DC (UL) |
| Bemessungsstrom | Rated current | 20A (2,5mm², 4mm²; 14AWG, 12AWG)<br>30A (6mm²; 10AWG)<br>43A (10mm²) |
| Prüfspannung | Test voltage | 6kV (50Hz, 1min.) |
| Umgebungstemperaturbereich | Ambient temperature range | -40°C...+90°C (IEC)<br>-40°C...+75°C (UL) |
| Obere Grenztemperatur | Upper limiting temperature | 105°C (IEC) |
| Schutzart, gesteckt<br>ungesteckt | Degree of protection, mated<br>unmated | IP67<br>IP2X |
| Überspannungskat./Verschmutzungsgrad | Overvoltage category/Pollution degree | CATIII/2 |
| Kontaktwiderstand der Steckverbinder | Contact resistance of plug connectors | 0,5mΩ |
| Schutzklasse | Safety class | II |
| Kontaktsystem | Contact system | MC Kontaktlamellen<br>MC Multilam |
| Anschlussart | Type of termination | Crimpen<br>Crimping |
| Kontaktmaterial | Contact material | Kupfer, verzinnt<br>Copper, tin plated |
| Isolationsmaterial | Insulation material | TPE/PA |
| Steckkraft/Auszugskraft | Insertion force/Withdrawal force | ≤ 50N/≥ 50N |
| Flammklasse | Flame class | UL94-HB/UL94-V0 |

 **Verschlusskappen** Seite 54
**Montagegeräte** Seite 58, 59

 **Sealing caps** page 54
**Assembly tools** page 58, 59

 Montageanleitung **MA207**
*www.multi-contact.com*

Assembly Instructions **MA207**
*www.multi-contact.com*

12

Courtesy of Muli-Contact USA

■ **Figure A-3**

# SUNNY BOY 5000US / 6000US / 7000US

> UL 1741/IEEE-1547 compliant

> 10 year standard warranty

> Highest CEC efficiency in its class

> Integrated load-break rated DC disconnect switch

> Integrated fused series string combiner

> Sealed electronics enclosure & Opticool™

> Comprehensive SMA communications and data collection options

> Ideal for residential or commercial applications

> Sunny Tower compatible

# SUNNY BOY 5000US/6000US/7000US
## The best in their class

Our US series inverters utilize our latest technology and are designed specifically to meet IEEE-1547 requirements. Sunny Boy 6000US and Sunny Boy 7000US are also compatible with the Sunny Tower. Increased efficiency means better performance and shorter payback periods. All three models are field-configurable for positive ground systems making them more versatile than ever. With over 750,000 fielded units, Sunny Boy is the benchmark for PV inverter performance and reliability throughout the world.

Courtesy of SMA Solar Technology AG

**Figure A-4** See detailed specifications at http://www.sma-america.com.

8

APPENDIX

## Technical Data
## SUNNY BOY 5000US / 6000US / 7000US

| | SB 5000US | SB 6000US | SB 7000US |
|---|---|---|---|
| Recommended Maximum PV Power (Module STC) | 6250 W | 7500 W | 8750 W |
| DC Maximum Voltage | 600 V | 600 V | 600 V |
| Peak Power Tracking Voltage | 250–480 V | 250–480 V | 250–480 V |
| DC Maximum Input Current | 21 A | 25 A | 30 A |
| DC Voltage Ripple | < 5% | < 5% | < 5% |
| Number of Fused String Inputs | 3 (inverter), 4 x 20 A (DC disconnect) | 3 (inverter), 4 x 20 A (DC disconnect) | 3 (inverter), 4 x 20 A (DC disconnect) |
| PV Start Voltage | 300 V | 300 V | 300 V |
| AC Nominal Power | 5000 W | 6000 W | 7000 W |
| AC Maximum Output Power | 5000 W | 6000 W | 7000 W |
| AC Maximum Output Current (@ 208, 240, 277 V) | 24 A, 21 A, 18 A | 29 A, 25 A, 22 A | 34 A, 29 A, 25 A |
| AC Nominal Voltage Range | 183 – 229 V @ 208 V<br>211 – 264 V @ 240 V<br>244 – 305 V @ 277 V | 183 – 229 V @ 208 V<br>211 – 264 V @ 240 V<br>244 – 305 V @ 277 V | 183 – 229 V @ 208 V<br>211 – 264 V @ 240 V<br>244 – 305 V @ 277 V |
| AC Frequency: nominal / range | 60 Hz / 59.3 - 60.5 Hz | 60 Hz / 59.3 - 60.5 Hz | 60 Hz / 59.3 - 60.5 Hz |
| Power Factor (Nominal) | 0.99 | 0.99 | 0.99 |
| Peak Inverter Efficiency | 96.8% | 97.0% | 97.1% |
| CEC Weighted Efficiency | 95.5% | 95.5% @ 208 V<br>95.5% @ 240 V<br>96.0% @ 277 V | 95.5% @ 208 V<br>96.0% @ 240 V<br>96.0% @ 277 V |
| Dimensions: W x H x D in inches | 18.4 x 24.1 x 9.5 | 18.4 x 24.1 x 9.5 | 18.4 x 24.1 x 9.5 |
| Weight / Shipping Weight | 141 lbs / 148 lbs | 141 lbs / 148 lbs | 141 lbs / 148 lbs |
| Ambient Temperature Range | −13 to 113 °F | −13 to 113 °F | −13 to 113 °F |
| Power Consumption: standby / nighttime | <7 W / 0.1 W | <7 W / 0.1 W | <7 W / 0.1 W |
| Topology | Low frequency transformer, true sinewave | Low frequency transformer, true sinewave | Low frequency transformer, true sinewave |
| Cooling Concept | OptiCool™, forced active cooling | OptiCool™, forced active cooling | OptiCool™, forced active cooling |
| Mounting Location: indoor / outdoor (NEMA 3R) | ●/● | ●/● | ●/● |
| LCD Display | ● | ● | ● |
| Lid Color: aluminum / red / blue / yellow | ●/○/○/○ | ●/○/○/○ | ●/○/○/○ |
| Communication: RS485 / wireless | ○/○ | ○/○ | ○/○ |
| Warranty: 10-year | ● | ● | ● |
| Compliance: IEEE-929, IEEE-1547, UL 1741, UL 1998, FCC Part 15 A & B | ● | ● | ● |

Specifications for nominal conditions

● Included  ○ Optional

### Efficiency Curves

www.SMA-America.com
Phone  916 625 0870
Toll Free  888 4 SMA USA

## SMA America, Inc.

**Figure A-5** See detailed specifications at http://www.sma-america.com.

SUNNY CENTRAL 250U / 500U

> 97% CEC weighted efficiency

> Integrated isolation transformer

> Graphical LCD interface

> Sunny WebBox compatible

> Optional combiner boxes

> Install indoors or out

> UL 1741 / IEEE-1547 compliant

# SUNNY CENTRAL 250U / 500U
## The ideal inverters for large scale PV power systems

The new Sunny Centrals have integrated isolation transformers and deliver the highest efficiencies available for large PV inverters. A completely updated user interface features a large LCD that provides a graphical view of the daily plant production as well as the status of the inverter and the utility grid. With the optional Sunny WebBox, users can now choose from either RS485 or Ethernet based communications. Designed for easy installation, operation and performance monitoring, the new Sunny Central is the ideal choice for your large scale PV project.

**Figure A-6** See detailed specifications at http://www.sma-america.com.

## Technical Data
## SUNNY CENTRAL 250U / 500U

| | SC 250U | SC 500U |
|---|---|---|
| Inverter Technology | True sine wave, high frequency PWM with galvanic isolation | True sine wave, high frequency PWM with galvanic isolation |
| AC Power Output (Nominal) | 250 kW | 500 kW |
| AC Voltage (Nominal) | 480 $V_{AC}$ WYE | 480 $V_{AC}$ WYE / Δ |
| AC Frequency (Nominal) | 60 Hz | 60 Hz |
| Current THD | < 5% | < 5% |
| Power Factor (Nominal) | > 0.99 | > 0.99 |
| AC Output Current Limit | 300 $A_{AC}$ @ 480 $V_{AC}$ | 600 $A_{AC}$ (@ 480 $V_{AC}$) |
| DC Input Voltage Range | 300 – 600 $V_{DC}$ | 300 – 600 $V_{DC}$ |
| MPP Tracking | 300 – 600 $V_{DC}$ | 300 – 600 $V_{DC}$ |
| PV Start Voltage (Configurable from 300 – 600 $V_{DC}$) | 400 $V_{DC}$ | 400 $V_{DC}$ |
| Maximum DC Current | 800 $A_{DC}$ | 1600 $A_{DC}$ |
| Peak Efficiency | 97.5% | 97.5% (estimated) |
| CEC Weighted Efficiency | 97% | 97% (estimated) |
| Power Consumption | 69 W Standby, <1000 W with fans | 69 W Standby, <1500 W with fans |
| Ambient Operating Temperature | −13 to 113 °F at full power output up to 122 °F at reduced power | −13 to 113 °F at full power output up to 122 °F at reduced power |
| Cooling | Variable-speed forced air | Variable-speed forced air |
| Enclosure | NEMA 3R | NEMA 3R |
| Dimensions: W x H x D in inches | 110 x 80 x 33 | 142 x 80 x 37 |
| Weight | 4200 lbs | 6725 lbs |
| Compliance | UL 1741, IEEE-1547 | UL 1741, IEEE-1547 (pending) |

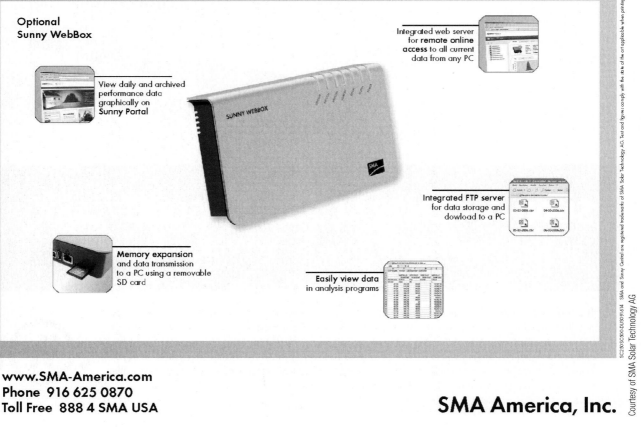

**Optional Sunny WebBox**

View daily and archived performance data graphically on **Sunny Portal**

Integrated web server for **remote online access** to all current data from any PC

Integrated FTP server for data storage and dowload to a PC

Memory expansion and data transmission to a PC using a removable SD card

**Easily view data** in analysis programs

**www.SMA-America.com**
**Phone  916 625 0870**
**Toll Free  888 4 SMA USA**

## SMA America, Inc.

■ **Figure A-7** See detailed specifications at http://www.sma-america.com.

## SUNNY BOY 5000US / 6000US / 7000US

> **UL 1741/IEEE-1547 compliant**

> **10 year standard warranty**

> **Highest CEC efficiency in its class**

> **Integrated load-break rated DC disconnect switch**

> **Integrated fused series string combiner**

> **Sealed electronics enclosure & Opticool™**

> **Comprehensive SMA communications and data collection options**

> **Ideal for residential or commercial applications**

> **Sunny Tower compatible**

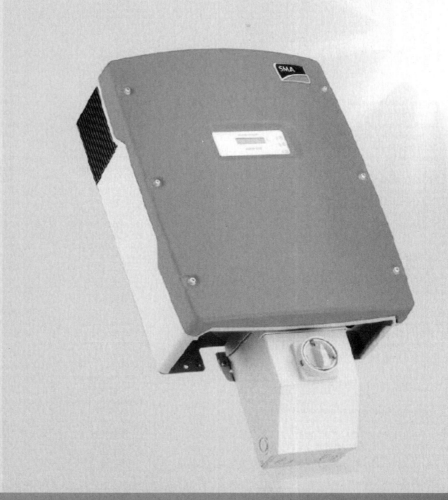

# SUNNY BOY 5000US/6000US/7000US
## The best in their class

Our US series inverters utilize our latest technology and are designed specifically to meet IEEE-1547 requirements. Sunny Boy 6000US and Sunny Boy 7000US are also compatible with the Sunny Tower. Increased efficiency means better performance and shorter payback periods. All three models are field-configurable for positive ground systems making them more versatile than ever. With over 750,000 fielded units, Sunny Boy is the benchmark for PV inverter performance and reliability throughout the world.

**Figure A-8** See detailed specifications at http://www.sma-america.com.

## Technical Data
## SUNNY BOY 5000US / 6000US / 7000US

| | SB 5000US | SB 6000US | SB 7000US |
|---|---|---|---|
| Recommended Maximum PV Power (Module STC) | 6250 W | 7500 W | 8750 W |
| DC Maximum Voltage | 600 V | 600 V | 600 V |
| Peak Power Tracking Voltage | 250–480 V | 250–480 V | 250–480 V |
| DC Maximum Input Current | 21 A | 25 A | 30 A |
| DC Voltage Ripple | < 5% | < 5% | < 5% |
| Number of Fused String Inputs | 3 (inverter), 4 x 20 A (DC disconnect) | 3 (inverter), 4 x 20 A (DC disconnect) | 3 (inverter), 4 x 20 A (DC disconnect) |
| PV Start Voltage | 300 V | 300 V | 300 V |
| AC Nominal Power | 5000 W | 6000 W | 7000 W |
| AC Maximum Output Power | 5000 W | 6000 W | 7000 W |
| AC Maximum Output Current (@ 208, 240, 277 V) | 24 A, 21 A, 18 A | 29 A, 25 A, 22 A | 34 A, 29 A, 25 A |
| AC Nominal Voltage Range | 183 – 229 V @ 208 V | 183 – 229 V @ 208 V | 183 – 229 V @ 208 V |
| | 211 – 264 V @ 240 V | 211 – 264 V @ 240 V | 211 – 264 V @ 240 V |
| | 244 – 305 V @ 277 V | 244 – 305 V @ 277 V | 244 – 305 V @ 277 V |
| AC Frequency: nominal / range | 60 Hz / 59.3 – 60.5 Hz | 60 Hz / 59.3 – 60.5 Hz | 60 Hz / 59.3 – 60.5 Hz |
| Power Factor (Nominal) | 0.99 | 0.99 | 0.99 |
| Peak Inverter Efficiency | 96.8% | 97.0% | 97.1% |
| CEC Weighted Efficiency | 95.5% | 95.5% @ 208 V | 95.5% @ 208 V |
| | | 95.5% @ 240 V | 96.0% @ 240 V |
| | | 96.0% @ 277 V | 96.0% @ 277 V |
| Dimensions: W x H x D in inches | 18.4 x 24.1 x 9.5 | 18.4 x 24.1 x 9.5 | 18.4 x 24.1 x 9.5 |
| Weight / Shipping Weight | 141 lbs / 148 lbs | 141 lbs / 148 lbs | 141 lbs / 148 lbs |
| Ambient Temperature Range | −13 to 113 °F | −13 to 113 °F | −13 to 113 °F |
| Power Consumption: standby / nighttime | <7 W / 0.1 W | <7 W / 0.1 W | <7 W / 0.1 W |
| Topology | Low frequency transformer, true sinewave | Low frequency transformer, true sinewave | Low frequency transformer, true sinewave |
| Cooling Concept | OptiCool™, forced active cooling | OptiCool™, forced active cooling | OptiCool™, forced active cooling |
| Mounting Location: indoor / outdoor (NEMA 3R) | ●/● | ●/● | ●/● |
| LCD Display | ● | ● | ● |
| Lid Color: aluminum / red / blue / yellow | ●/○/○/○ | ●/○/○/○ | ●/○/○/○ |
| Communication: RS485 / wireless | ○/○ | ○/○ | ○/○ |
| Warranty: 10-year | ● | ● | ● |
| Compliance: IEEE-929, IEEE-1547, UL 1741, UL 1998, FCC Part 15 A & B | ● | ● | ● |

Specifications for nominal conditions                    ● Included    ○ Optional

**Efficiency Curves**

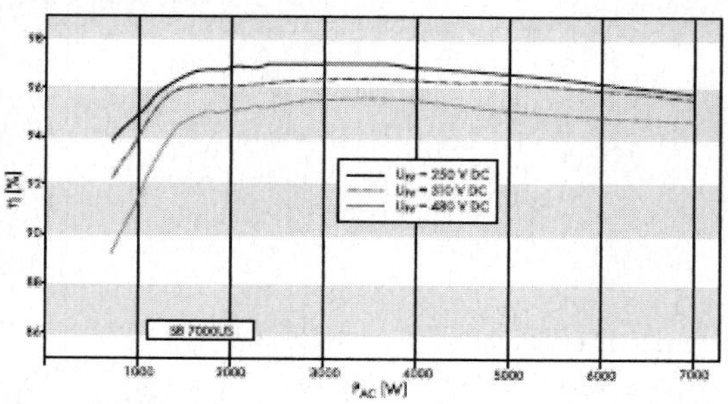

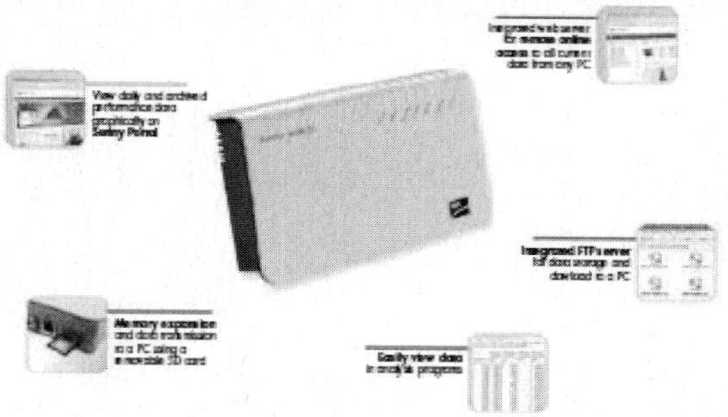

www.SMA-America.com
Phone  916 625 0870
Toll Free  888 4 SMA USA

**SMA America, Inc.**

Courtesy of SMA Solar Technology AG

**Figure A-9** See detailed specifications at http://www.sma-america.com.

## SUNNY SENSORBOX

### Precise
> Complete system monitoring for PV-plants
> Measurement of insolation, module temperature, ambient temperature and wind speed

### Easy to Use
> Easy installation right next to the solar modules
> Integration into existing systems via serial interface RS485
> Compatible with Sunny WebBox and Sunny Boy Control
> Data analysis using a PC or in the Sunny Portal

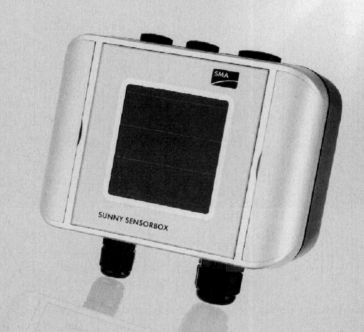

# SUNNY SENSORBOX
## Perfect performance monitoring

It is one of the smallest measurement units available and extremely easy to install: the new Sunny SensorBox from SMA. This has been specially developed to improve PV-system performance analysis even further – at an attractive price. The Sunny SensorBox now lets you acquire data concerning the ambient conditions such as solar irradiation and module temperature in order to give you the ability to detect malfunctions or yield losses as early as possible. With this we satisfy the demands from PV-plant operators for perfect plant performance monitoring – and make a further contribution towards optimal yield security.

■ **Figure A-10** See detailed specifications at http://www.sma-america.com.

# SUNNY SENSORBOX

**Innovation and precision for your performance monitoring**

### Complete system monitoring easily installed

The Sunny SensorBox is installed outdoors at the solar generator, and comes with an integrated solar cell, which measures solar irradiation. The module temperature is measured by means of the temperature sensor which is included. From the present solar irradiation level and the module temperature, it is possible to calculate the expected output, and to compare it with the actual measured output of the inverters. Temporary or continuous yield losses caused by unknown failure sources are therefore a thing of the past.

### ... extendable

Once the Sunny SensorBox has been equipped according to the modules, it is simply connected with the inverter to a Sunny WebBox with an RS485 data connection. From there, the data can be transferred to a PC for further processing, or to the Sunny Portal for automatic performance analysis. The Sunny SensorBox also enables the connection of additional sensors, e. g. to measure the ambient temperature or wind speed for calculations which are even more precise. This ensures reliable system monitoring for operators – and maximum yield security.

**■ Figure A-11** See detailed specifications at http://www.sma-america.com.

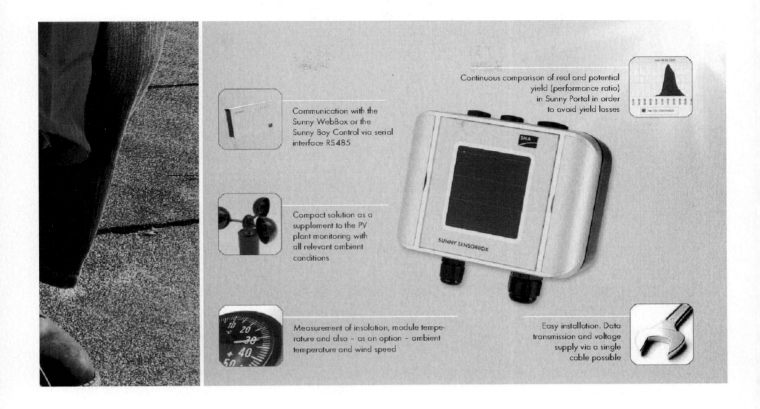

Communication with the Sunny WebBox or the Sunny Boy Control via serial interface RS485

Continuous comparison of real and potential yield (performance ratio) in Sunny Portal in order to avoid yield losses

Compact solution as a supplement to the PV plant monitoring with all relevant ambient conditions

Measurement of insolation, module temperature and also – as an option – ambient temperature and wind speed

Easy installation. Data transmission and voltage supply via a single cable possible

### Performance ratio as a quality indicator

Shadowing, defects, surface contamination and gradual malfunctions such as deteriorating modules have a serious impact on the generator yield and the overall performance and are not to be underestimated. Particularly annoying for the operator is the fact that the losses in yield could have been avoided in most cases – if the error had been detected in time. The system efficiency of the PV-plant (performance ratio) is therefore an essential value. The performance ratio indicates the ratio of actual yield to the theoretically possible yield. Since the performance ratio indicates how the irradiated energy on the generator side is exploited, it is the decisive quality factor for the performance of the entire PV system. This is where the Sunny SensorBox comes into play.

### How to determine the performance ratio

You simply divide the actual energy yield through the possible energy yield. While the inverter measures the actual energy, the possible energy yield is determined according to the efficiency of the modules, the module surface and the recorded insolation. Good grid connected PV systems reach performance ratios of between 60 % and 80 % – ratios under this value can indicate malfunctions of the system.

**Figure A-12** See detailed specifications at http://www.sma-america.com.

# Technical Data
## SUNNY SENSORBOX

| | Sunny SensorBox | |
|---|---|---|
| **Interfaces** | | |
| to the data logger (Sunny WebBox, Sunny Boy Control) | RS485 | |
| **Internal sensor** | | |
| Solar irradiation | ASI solar module, Precision ±8 % Range 0 ... 1500 W/m² | |
| **External sensor** | | |
| Module temperature | Platinum Sensor (Pt100) attachable Range –20 °C ... +110 °C Precision ±0.5 °C | |
| **Optional sensors** | | |
| Ambient temperature | Platinum Sensor (Pt100) Range –30 °C ... +80 °C Precision ±0.5 °C | |
| Wind measurement | Thies Clima external anemometer Range 0.8 m/s ... 40 m/s (max. 60 m/s short term) Precision ±0.5 m/s | |
| **Power supply** | | |
| via RS485 line | Via external power supply (Power Injector) | |
| **Protection rating** | | |
| in accordance with DIN EN 60529 | IP65 | |
| **Mechanical data** | | |
| Width / height / depth (mm) | 120 / 90 / 50 | |
| Weight | 500 g | |

### Schematic diagram of Sunny SensorBox

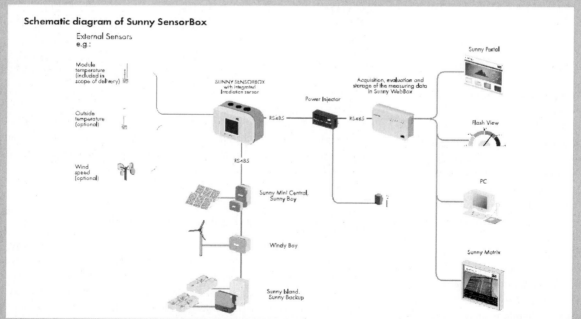

SENSORBOX-DEN-083124  SMA and Sunny Boy are registered trademarks of SMA Solar Technology AG. Text and illustrations reflect the technical state at the time of printing. Technical modifications reserved. No liability accepted for printing errors. Printed on chlorine-free paper.

Courtesy of SMA Solar Technology AG

**www.SMA.de**
**Freecall +800 SUNNYBOY**
Freecall +800 78669269

**SMA Solar Technology AG**

■ **Figure A-13** See detailed specifications at http://www.sma-america.com.

# SUNNY CENTRAL STRING-MONITOR US

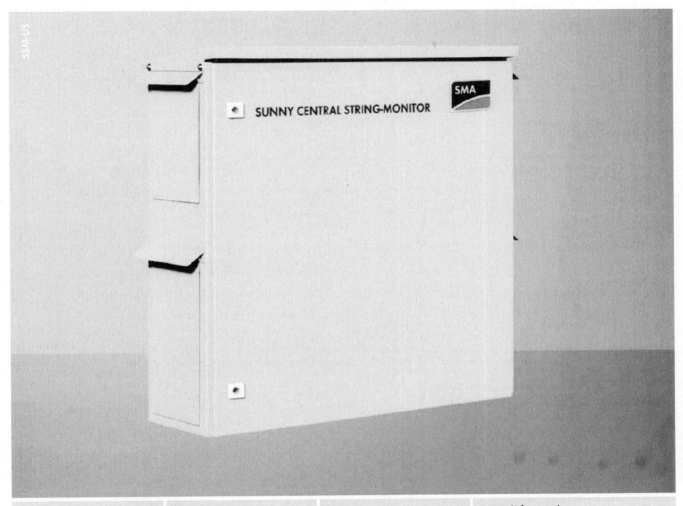

### Efficient
- Optimum fault recognition for high yields
- Reduced system costs through use of up to nine devices per Sunny Central inverter

### Precise
- NEMA 3R DC distributor box with integrated current measurement

### Flexible
- Three models for the best possible system design
- Optionally suitable for use near the coast

### Straightforward
- Removable floor and side plates allow for easy mounting
- Simple configuration of string current monitoring

## SUNNY CENTRAL STRING-MONITOR US

Accurate monitoring for large North American systems ensures high yields

Detailed string cluster monitoring: the Sunny Central String-Monitor US. By measuring and comparing individual string currents, power deviations in the solar array are reliably detected and analyzed in the Sunny Central. The Sunny Central String-Monitor US is available in three models, each with varying NEC-compliant string fuse protection, making the Sunny Central String-Monitor US well suited for all panel types in the North American solar market.

■ **Figure A-14** See detailed specifications at http://www.sma-america.com.

# Technical Data

| | Sunny Central String-Monitor US Version 24 | Sunny Central String-Monitor US Version 32 | Sunny Central String-Monitor US Version 64 |
|---|---|---|---|
| **PV generator connection** | | | |
| Input voltage range | 0 ... 600 V DC | 0 ... 600 V DC | 0 ... 600 V DC |
| Max fuse size (10 x 38 class CC fuses) | 20 A, 600 V DC 1) | 15 A, 600 V DC 1) | 8 A, 600 V DC 1) |
| Max. PV short-circuit current per string | 12.8 A 2) | 9.6 A 2) | 5.1 A 2) |
| Max. number of strings | 24 | 32 | 64 |
| Fused inputs per measuring channel | 3 | 4 | 8 |
| PV array configuration | neg. or pos. grounded | neg. or pos. grounded | neg. or pos. grounded |
| Number of measuring channels | 8 | 8 | 8 |
| **Sunny Central Connection** | | | |
| DC short-circuit current | 480 A | 480 A | 512 A |
| Max. operating output current, A DC, continuous | 308 A | 308 A | 328 A |
| Max. number of cables per output port | 2 | 2 | 2 |
| **Mechanical Data** | | | |
| Dimensions: W x H x D in inches | 31.5 x 31.5 x 9.8 | 31.5 x 31.5 x 9.8 | 31.5 x 47.2 x 11.8 |
| Weight | 143 lbs | 146 lbs | 194 lbs |
| Protection rating | NEMA 3R | NEMA 3R | NEMA 3R |
| Housing material | Steel or Aluminum | Steel or Aluminum | Steel or Aluminum |
| **Ambient conditions** | | | |
| Permissible ambient temperatures | –13 °F to 113 °F | –13 °F to 113 °F | –13 °F to 113 °F |
| Rel. humidity | up to 95%, condensation possible | up to 95%, condensation possible | up to 95%, condensation possible |
| **Communication** | | | |
| Connection SSM-US | RS485 | RS485 | RS485 |

1) These values indicate input current without the derating factor of NEC articles 690.8(A)(1) and 690.8 (8)(1) applied.
2) These values indicate input current with the derating factor of NEC articles 690.8(A)(1) and 690.8 (8)(1) applied.

| Type designation | SSM-US | SSM-US | SSM-US |
|---|---|---|---|

## SSM-US Version 32, Negative Grounded

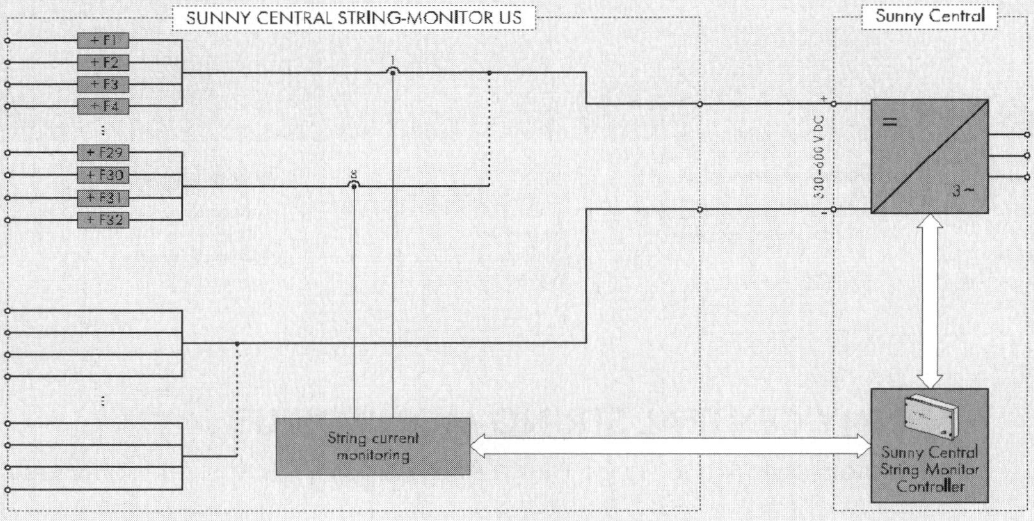

Tel. +1 916 625 0870
Toll Free +1 888 4 SMA USA
www.SMA-America.com

**SMA America, LLC**

SUNNYSTRINGUSUS100716 SMA and Sunny Central are registered trademarks of SMA Solar Technology AG. Text and figures comply with the state of the art applicable when printing. Subject to technical changes. We accept no liability for typographical and other errors. Printed on chlorine-free paper.

Courtesy of SMA Solar Technology AG

■ **Figure A-15** See detailed specifications at http://www.sma-america.com.

# HIT Photovoltaic Module

 Power 215N

Photovoltaic Module

**Module Efficiency: 17.1%**
**Cell Efficiency: 19.3%**
**Power Output - 215 Watts**

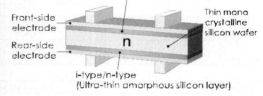

### SANYO HIT® Solar Cell Structure

p-type/i-type
(Ultra-thin amorphous silicon layer)

Front-side electrode

Rear-side electrode

n

Thin mono crystalline silicon wafer

i-type/n-type
(Ultra-thin amorphous silicon layer)

### SANYO'S Proprietary Technology

HIT solar cells are hybrids of mono crystalline silicon surrounded by ultra-thin amorphous silicon layers, and are available solely from SANYO.

### High Efficiency

HIT® Power solar panels are leaders in sunlight conversion efficiency. Obtain maximum power within a fixed amount of space. Save money using fewer system attachments and racking materials, and reduce costs by spending less time installing per watt. HIT Power models are ideal for grid-connected solar systems, areas with performance based incentives, and renewable energy credits.

### Power Guarantee

SANYO's power ratings for HIT Power panels guarantee customers receive 100% of the nameplate rated power (or more) at the time of purchase, enabling owners to generate more kWh per rated watt, quicken investments returns, and help realize complete customer satisfaction.

### Temperature Performance

As temperatures rise, HIT Power solar panels produce 10% or more electricity (kWh) than conventional crystalline silicon solar panels at the same temperature.

### Valuable Features

The packing density of the panels reduces transportation, fuel, and storage costs per installed watt.

### Quality Products Made in USA

SANYO silicon wafers located inside HIT solar panels are made in California and Oregon (from October 2009), and the panels are assembled in an ISO 9001 (quality), 14001 (environment), and 18001 (safety) certified factory. Unique eco-packing minimizes cardboard waste at the job site. The panels have a Limited 20-Year Power Output and 5-Year Product Workmanship Warranty.

### Unnecessary Section When Using SANYO

### Increased Performance with SANYO

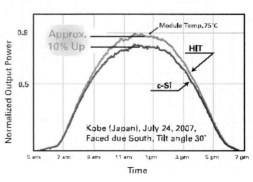

Courtesy of Sanyo North America Corporation

■ **Figure A-16** See us.sanyo.com for complete specifications.

## HIT Power 215N
Photovoltaic Module

### Electrical Specifications

| Model | HIT Power 215N or HIP-215HKHA6 |
|---|---|
| Rated Power (Pmax)[1] | 215 W |
| Maximum Power Voltage (Vpm) | 42.0 V |
| Maximum Power Current (Ipm) | 5.13 A |
| Open Circuit Voltage (Voc) | 51.6 V |
| Short Circuit Current (Isc) | 5.61 A |
| Temperature Coefficient (Pmax) | -0.336%/ °C |
| Temperature Coefficient (Voc) | -0.143 V/ °C |
| Temperature Coefficient (Isc) | 1.96 mA/ °C |
| NOCT | 114.8°F (46°C) |
| CEC PTC Rating | 199.6 W |
| Cell Efficiency | 19.3% |
| Module Efficiency | 17.1% |
| Watts per Ft.[2] | 15.85 W |
| Maximum System Voltage | 600 V |
| Series Fuse Rating | 15 A |
| Warranted Tolerance (-/+) | -0% / +10% |

### Mechanical Specifications

| Internal Bypass Diodes | 3 Bypass Diodes |
|---|---|
| Module Area | 13.56 Ft² (1.26m²) |
| Weight | 35.3 Lbs. (16kg) |
| Dimensions LxWxH | 62.2x31.4x1.8 in. (1580x798x46 mm) |
| Cable Length +Male/-Female | 40.55/34.6 in. (1030/880 mm) |
| Cable Size / Connector Type | No. 12 AWG / MC4™ Locking Connectors |
| Static Wind / Snow Load | 60PSF (2880Pa) / 39PSF (1867Pa) |
| Pallet Dimensions LxWxH | 63.2x32x72.8 in. (1607x815x1850 mm) |
| Quantity per Pallet / Pallet Weight | 34 pcs./1234.5 Lbs (560 kg) |
| Quantity per 53' Trailer | 952 pcs. |

### Operating Conditions & Safety Ratings

| Ambient Operating Temperature | -4°F to 115°F (-20°C to 46°C)[2] |
|---|---|
| Hail Safety Impact Velocity | 1" hailstone (25mm) at 52 mph (23m/s) |
| Fire Safety Classification | Class C |
| Safety & Rating Certifications | UL 1703, cUL, CEC |
| Limited Warranty | 5 Years Workmanship, 20 Years Power Output |

[1]STC: Cell temp. 25°C, AM1.5, 1000W/m² [2]Monthly average low and high of the installation site.
Note: Specifications and information above may change without notice.

### Dimensions
Unit: inches (mm)

### Dependence on Temperature

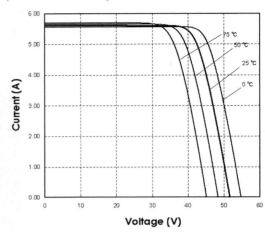

### Dependence on Irradiance

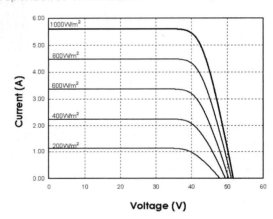

⚠ **CAUTION!**

Read the operating instructions carefully before use of these products

## SANYO

**SANYO Energy (U.S.A.) Corp.**
A Division of SANYO North America Corporation

550 S. Winchester Blvd., Suite 510
San Jose, CA 95128, U.S.A.
www.sanyo.com/solar
solar@sec.sanyo.com

■ **Figure A-17** See us.sanyo.com for complete specifications.

# INDEX

Page numbers followed by "*f*" indicate figure; and those followed by "*t*" indicate a table.